3-02

Landscape and Atmosphere

Certificate Geography

by
John G. Wilson B.A.

 Schofield & Sims Ltd   Huddersfield

0 7217 1031 X

0 7217 1037 9 Net edition

First printed 1972

Second impression 1973

Third impression 1974

Fourth impression 1975

Fifth impression 1976

Sixth impression 1977

Seventh impression 1977

Eighth impression 1978

Ninth impression 1979

Tenth impression 1980

Eleventh impression 1982

Certificate Geography is a series of five books:

Landscape and Atmosphere	0 7217 1031 X
Man in his Environment	0 7217 1033 6
World of Man	0 7217 1032 8
Language of Maps	0 7217 1034 4
The British Isles	0 7217 1038 7

Printed in England by Netherwood Dalton & Co Ltd Huddersfield

Contents

Acknowledgements

The author and publishers wish to acknowledge the Examining Boards and photographic sources and thank the owners for permission to use their copyright material appearing on the following pages:

National Aeronautics and Space Administration *7*

Crown Copyright Geological Survey photograph. Reproduced by permission of the Controller of H.M. Stationery Office *12*

J. G. Durnall *12, 35, 54, 57, 61*

Miss E. Anderton *13*

D. W. A. Laurence *19*

B. F. D. Harris *19, 22, 37*

Swiss National Tourist Office *22, 52*

Northern Ireland Tourist Board *26*

Japan Information Centre *26*

T. Wanless *28, 98*

Dr. R. J. Rice *30, 34, 55, 74*

Ruth Mason, Kindrogan Field Centre *34*

Miss J. Berry *37, 63*

P. Hitt *39, 46, 47, 53*

Aerofilms Limited *49, 52, 58, 61, 62, 64*

W. L. Smith, Ph.D. *51*

J. A. Jackson, Plas y Brenin National Mountaineering Centre *52*

A. S. Norsk Telegrambyra *65*

Australian News and Information Bureau *67*

Shell International Petroleum Co. Ltd. *70*

W. F. Davidson, Penrith *73*

J. K. Vose *74*

Water Resources Board *74*

J. E. L. Hulbert *82, 83, 84*

J. T. Bartlett *97*

A. Tabor *98 and cover*

W. G. Pendleton *104*

Examining Boards *106 - 109*

Preface

This volume covers the topics that are generally included in the study of physical geography. It provides the reader with the information needed to give full and meaningful answers to the questions commonly set at the lower level of public examinations. The many photographs have been closely integrated into the text and, wherever possible, show locations within the British Isles. This is designed to give a greater sense of reality to the subject, and to encourage the reader to seek out the examples that will doubtless be found, if not on his doorstep, at least within a short distance of his home.

Above all, it is hoped that this book will be read with pleasure, and will prompt in the reader a greater awareness of the fascination of landscape and atmosphere.

I am much indebted to those who kindly allowed me to use their colour transparencies and to the many people who assisted in tracking down suitable illustrations. I would also like to express my thanks to Ken Vose for help with problems of metrication, and to Joan Wood without whose long and charitable memory this book would never have been started.

J.G.W.

Fig 1

Introduction

It is appropriate that a photograph is our starting point, for Geography is a subject greatly concerned with what we can see about us. This particular view, which shows a landscape in east Yorkshire, was taken on a bright day in June when the atmosphere was clear and still. It is a placid scene and one imagines it to be unchanging. In this it is most deceptive, for atmosphere and landscape are each subject to many and varied processes that bring about change. It is these processes and these changes that are the theme of this book.

It is in the state of the atmosphere that change may most readily be appreciated. A sudden shower sends us scurrying for shelter, and a break in the drifting clouds gives us a welcome burst of sunshine. Simple instruments reveal other variations. The weathercock on your local church is sensitive to slight shifts of wind direction and the thermometer records the rise and fall of air temperature.

The land surface is also undergoing constant modification. Look again at the photograph. Solid rock underlies the vegetation and the soil in which it grows. This rock, once the bed of a long-vanished sea, is under constant attack by natural agencies. It is slowly being broken down into smaller and smaller fragments by a variety of processes known collectively as *weathering*. The weathering of the solid rock facilitates the work of the agents of *erosion*. Weathering and erosion work hand in hand and their combined activities are known as *denudation*. Together they carve and shape the land surface and generally wear it away. Acting on rocks of different hardness they have carved the varied landscape that we see in the photograph.

The river in the foreground is an example of an agent of erosion. It is the River Derwent and it is slowly wearing away its channel as it flows along. Another example may be glimpsed through the gap between the steep hillsides. This is the North Sea, and breaking waves are constantly nibbling away at the steep cliffs of this part of the Yorkshire coast. The destructive effects of glaciers and ice-sheets that were active during the distant Ice Age can often be recognised in the present day landscape. In some areas, the wind achieves the power and status of an agent of erosion.

Agents of erosion have other tasks to perform. As the Derwent winds along, it is transporting a load of rock fragments derived from weathering and its own erosive work. Later, it will deposit this material either on the land to build up a new surface, or in the sea to form the raw material of future rocks. Sea, wind and ice also transport and deposit, and thus further modify the land surface. A small example may be detected in the photograph. During the Ice Age a tongue of ice pushed inland through the gap. It has left its mark on the landscape in the low ridge of deposited material that may be identified on the extreme right of the photograph.

The processes mentioned above operate so slowly that change is usually imperceptible. Only occasionally, as in the case of a landslide or destructive flood, is it sudden and dramatic. Nevertheless, on a country walk or a sea-side holiday we can look at the landscape and recognise the processes at work and identify the tell-tale signs of their constant activity.

Atmosphere and landscape are of more than academic interest. They greatly influence the life and activity of man. The farmer, for instance, always keeps a weather-eye on the state of the atmosphere – and with good reason. Wet weather at haymaking time means hard work and financial loss. A late frost reduces his chances of a profitable fruit crop.

The form of the land surface, its height and slope, also has a great influence on man. This may be illustrated from the photograph. The farm that has been built on the ridge deposited by ice enjoys safety from flooding in this ill-drained area. The course of the road has been influenced by similar considerations. Note how the steepest slopes are left in woodland when all other land has been cleared for agriculture. The photograph also contains a reminder that man is not powerless in face of this influence. The feature that cuts diagonally across the foreground is a large drainage channel that takes excess water from the Derwent through the gap to the sea. This reduces the risk of flooding both here and in areas further downstream.

Geography has much to say about the relationship between man and the world around him. An understanding of landscape and atmosphere is essential to a sound appreciation of this relationship.

1
The
Nature of
the Earth

The crust in motion

The photograph below was taken on the historic Apollo 11 mission to the moon. It is an astronaut's-eye view of the earth from a distance of about 160 000 km. Most of Africa and the Middle East can readily be identified, but elsewhere the surface of the earth is obscured by swirling masses of cloud.

At a time when man is rapidly expanding his knowledge of space, it is as well to remember that he has only a limited knowledge of the interior of his own planet. Man may have travelled the 376 285 km to the moon, but the furthest he has gone towards the centre of the earth is 3428 m to the bottom of a South African gold mine. Even the deepest hole he has bored, an American oil well 7742 m deep, must be considered the merest scratch on the surface when compared with the earth's radius of 6371 km.

Fig 2

Fortunately, the earth itself provides clues to the nature of its own interior. Earthquakes send out different types of shock waves which travel at different speeds according to the nature of the material they pass through. They are recorded by sensitive seismographs all over the world. From a study of these earthquake waves, scientists have built up a picture of the interior of the earth.

It is composed of a series of concentric shells which differ in composition and density. The *core* of the earth, approximately half its radius, is a dense mass of nickel and iron. This is wrapped in a *mantle* composed of progressively less dense material. The mantle is capped by a thin, light *crust* which is shown diagrammatically in Fig 3.

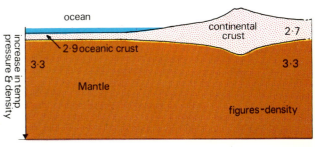

Fig 3

The oceanic crust is approximately 8 km thick, but that of the continental masses averages 20 km, and achieves twice this figure in areas of high mountains. Both types of crust rest on denser mantle material. Temperature increases rapidly towards the centre of the earth. This is due to radio-activity, and the rate of increase is such that great heat is often a problem in deep mines. Scientists estimate that temperatures within the upper mantle may reach thousands of degrees Celsius. Such temperatures are high enough to turn any rock into liquid. This does not normally happen because of the tremendous pressure exerted by the great thicknesses of overlying material.

However, certain layers within the upper mantle have a degree of plasticity. They are able to 'flow' – but movement is exceedingly slow.

The crust of the earth is not a continuous skin like the peel of an orange. It is broken into a number of rigid sections called *plates*. The largest of these are indicated and named in Fig 4. These plates are seldom still. Driven, it is believed, by convection currents seated deep in the earth's interior, they creep over the plastic mantle. Individual plates move in different directions, and their interaction helps to explain important features of the earth's crust. The plates themselves are seldom disturbed, but their fringes are the scenes of slow but powerful changes. Oceans expand, mighty mountain ranges are uplifted and rocks from the interior are piled on the surface in volcanoes. Earthquakes are the vibrant signs of plates in motion. Plate boundaries are mapped by plotting the point of origin of a large number of these disturbances.

Plates seldom move at speeds of more than 5 centimetres a year. Time, however, is something the geologist has in abundance. The earth is believed to be about $4\frac{1}{2}$ thousand million years old, and with a time scale like this, even the slowest processes can produce significant results. The effects of plate movements are varied and are best illustrated by examples.

Along a line which roughly bisects the Atlantic Ocean, the American plate is moving away from those of Eurasia and Africa. Molten mantle material rises up to fill the gap, and solidifies into solid rock – the new floor of an expanding ocean. A submarine ridge of considerable elevation is created by this volcanic activity.

The Pacific and China plates are in head-on conflict. The former dips down beneath its adversary, and is slowly absorbed in the hot interior. As the plate sinks it drags the ocean floor into a deep narrow trough. The downward movement is not smoothly achieved, and friction leads to frequent earthquakes and much volcanic activity.

The boundary between the Pacific and American plates runs through coastal California. The Pacific plate is moving NNW relative to its neighbour at a speed of 5 centimetres a year. Occasionally, movement may be seriously retarded by friction. This causes a build up of pressure which can only be relieved by a major earthquake. Such an earthquake destroyed San Francisco in 1906. Another is possible at any time.

Rocks of the earth's crust

The continental crust is composed of rocks in great variety. Fig 5 illustrates just a handful of the many different types, and other examples may be seen about us, for many rocks are of economic value. The slate on the roofs of older houses, the chippings used to surface minor roads, and the usual contents of a coal scuttle are

Fig 4

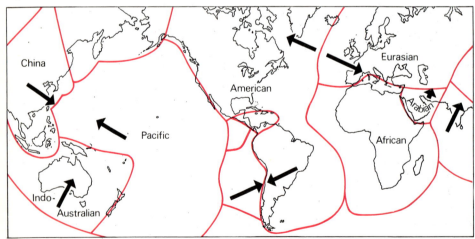

China
Pacific
Indo-
Australian
American
Eurasian
Arabian
African

⎯⎯ plate boundaries

➜ direction of movement

Fig 5

pressure is locally reduced. It is composed of various minerals each with its own particular chemical composition. Magma may be made up of different combinations of minerals and so igneous rocks show a wide range of chemical composition. As magma cools its minerals crystallise. If it cools quickly, as it does when it appears on the surface as lava, there is little time for crystals to develop and they are often invisible to the naked eye. Such rocks are termed *volcanic*. If magma cools slowly beneath the surface, the crystals are large and the rock is known as *plutonic*. This type of igneous rock is only revealed when denudation has removed the overlying rock. Granite is, perhaps, the best known example of a plutonic rock. A specimen from Shap in Westmorland is shown in Fig 5 (iii). The large crystals can easily be identified. The large pink crystals are of feldspar, silica is the white, and the mica appears as black flakes. Basalt, a common type of volcanic rock, forms the steep cliffs in Fig. 37.

Igneous rocks are hard and durable, and hence are much used for road material. The full beauty of the crystalline pattern of many plutonic varieties is revealed by polishing, and they are favoured as a decorative building stone. Many examples may be identified in the public buildings of most towns, and as headstones in churchyards.

Sedimentary rocks are formed by the cementing together of accumulations of small particles of material produced by geological processes. The material may be small fragments of older rocks, in which case the term *detrital* is used; it may be the remains of animal or plant life; or it may accumulate as a result of chemical processes.

We can frequently observe the early stages in the formation of detrital rocks. Fig 6 illustrates a common situation around our coasts. The river flowing down to the sea is carrying a load of rock fragments of various sizes picked up on its journey across the land surface. The load is deposited in the sea, but sorted according to size. The heavier pebbles are deposited first, then the sand, and finally the finest mud. The sand of the holiday beach is the

only a few examples. Your local museum will doubtless have a display of a wide range of rock samples that will repay examination and perhaps your school has a representative selection. Best of all, try to build up your own collection. You will find exposures of rock wherever the soil cover has been removed. Cuttings, quarries and along the banks of rivers are good places to look, but always have a sensible regard for your own safety. By examining the rock where it occurs and by handling specimens you will be impressed by its great variety, not only in appearance and colour, but also in composition and texture.

Rocks are classified according to the way in which they were formed.

Igneous rocks are formed by the cooling of molten material. This material, known as *magma*, is derived from the areas of high temperature beneath the surface where

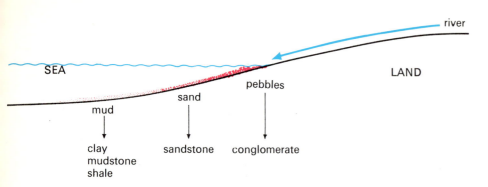

SEA

LAND

river

pebbles

sand

mud

clay
mudstone
shale

sandstone   conglomerate

Fig 6

raw material of future sandstone; the mud that covers the sea floor beyond will become, in the course of time, clay, mudstone or shale. The sand and mud particles will be compressed by the weight of overlying material and cemented together by minerals such as compounds of iron deposited by water circulating between the grains. Similarly, pebbles are cemented together to form the rock known as *conglomerate*.

The composition of these rocks depends on the nature of the material deposited. Their colour depends largely upon the minerals that form the cement. Sandstones, for instance, occur in a rainbow of colours from red, through brown and tan, to a delicate shade of green. It is solely particle size that determines the type of rock.

Limestone, which is mainly composed of calcium carbonate, is the most common rock formed from the remains of living organisms. The seas are alive with creatures, often microscopic in size, that extract calcium carbonate from sea water in order to build up the hard parts of their bodies. The remains of these creatures accumulate on the floor of the sea to be cemented into limestone with the passage of time. The massive limestone that today forms much of the Pennines was formed in this way in seas that existed over 300 million years ago. Here and there, the rock is composed of the remains of particular forms of marine life. Areas where shellfish were predominant we recognise today as shelly limestone.

Fig 5 (iv) shows a polished sample of limestone that is composed of the tightly packed remains of a sea creature known as a crinoid, which flourished in these ancient seas.

Chalk is a pure, fine-grained limestone. It underlies much of South-East England and is responsible for the whiteness of the cliffs of Dover.

That rocks may be formed from the remains of plant life is shown by the existence of the black and brittle rock we know as coal.

An example of a sedimentary rock produced by chemical processes is the great belt of *oolitic* limestone that extends from the south coast, through the Cotswolds, to North-East Yorkshire. This limestone is due to the fact that, under certain conditions, calcium carbonate is precipitated directly from sea water to form tiny grains that accumulate on the sea floor. It provides an attractive building stone and is, in some areas, a valuable source of iron ore.

If a solution of brine is allowed to evaporate, crystals of common salt will be deposited on the bottom of the container. Nature does this on a gigantic scale. The process can be observed today in many salt lakes in the arid parts of the world, such as Australia or Western U.S.A. The presence of thick beds of salt in the rocks which underlie much of Mid-Cheshire, for example, is evidence that this process operated in former times. Gypsum and potash are other chemical rocks of economic importance that were formed by evaporation.

Metamorphic rocks are rocks which have been changed as a result of the high temperatures and intense pressures present within the earth's crust. The slate of North Wales is a good example. Originally this rock was a soft mudstone, but it was metamorphised into the hard rock that splits so conveniently into roofing material. Marble is a metamorphic rock derived from limestone. Even hard igneous rock can be changed. Granite, for example, may be transformed into gneiss.

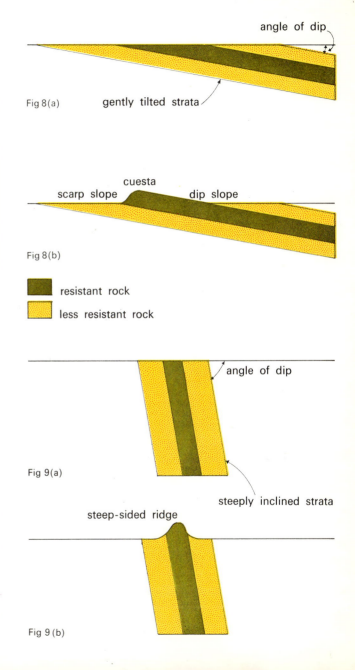

Fig 8(a)   angle of dip

gently tilted strata

Fig 8(b)   scarp slope   cuesta   dip slope

resistant rock

less resistant rock

Fig 9(a)   angle of dip

steeply inclined strata

Fig 9(b)   steep-sided ridge

Fig 7

The arrangement of the rocks

The quarry in Fig 7 shows an exposure of sedimentary rock. The horizontal lines visible on the face of the rock are known as *bedding planes*. They divide the rock into *strata* (sing. *stratum*) and represent pauses in the process of deposition when the material was being accumulated. The same term also describes the division between different types of sedimentary rock. The short vertical cracks are known as *joints*, and these, unlike bedding planes, are also commonly found in igneous and metamorphic rocks.

The strata in this exposure are virtually horizontal. This is what we would expect when we consider that the rocks were formed on the generally level floor of the sea. But many things can happen to rocks during the long eras of geological time. Earth movements, for instance, may tilt the strata to any angle. In Fig 8 (a) the tilt, or angle of dip, is only a matter of a few degrees, but, as (b) shows, when denudation has done its work the familiar land-form of the cuesta is formed. Steeply inclined rocks may lead to the formation of a steep-sided ridge or 'hog's back', Fig 9.

Fig 10

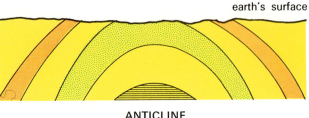

earth's surface

ANTICLINE

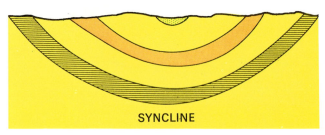

SYNCLINE

Fig 12

The rocks in Figs 10 and 11 are sedimentary rocks, formed in level layers, but look at them now. Movement of the earth's crust has folded them as though they were paper. The two commonest types of *folds* are named in Fig 12.

Look now at Fig 13, which again shows an exposure of sedimentary rock. The line that crosses the rock diagonally is a great crack or *fault*. By comparing the strata on each side of the fault we can see that the rock on the left of the photograph has slipped down relative to that on the right. The amount of slip or 'throw' is only a matter of a few metres in this example, but a major fault may have a throw of thousands of metres. It is movement of part of the earth's crust along a fault that gives rise to the vibrations we know as earthquakes.

Fig 11

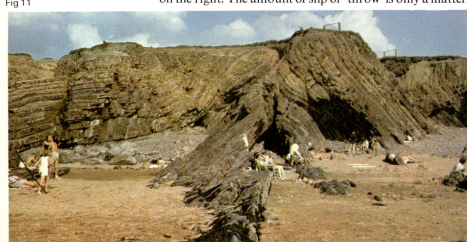

Fig 13

The examples given have been of sedimentary rocks, but it must be stressed that all types of rock may be tilted, folded or faulted. Fig 14 shows metamorphic rock that has suffered both folding and faulting.

Fig 14

The age of the rocks

The radio-active elements, such as uranium, contained in rocks change into other elements at a fixed rate. This gives the geologist a 'clock' that can be used to calculate the age of a rock sample with a high degree of accuracy. Using this method it has been found that the oldest known rocks were formed over 3 500 000 000 years ago. Geological time is divided into eras and periods which are given, with their dates, in Fig 15.

The rock exposed in the quarry shown in Fig 7 belongs to the Jurassic period, which, as Fig 15 shows, lasted from 195 to 135 million years ago. At that time the area of the quarry was part of the floor of an extensive sea. On the floor of this sea, calcium carbonate was slowly accumulating to great thicknesses, later to be converted into the oolitic limestone which is being quarried today. This Jurassic sea was home for a variety of marine organisms. As they died, the hard parts of their bodies resisted decay and were incorporated in the rock to be preserved as fossils. Examples are shown in Fig 16.

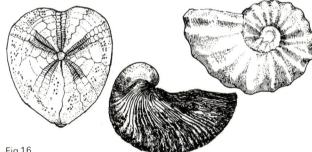

Fig 16

Fig 15

Era	Periods	Millions of years ago	Examples of Rock Formations in Britain	Mountain Building Periods	Life
Quaternary	Pleistocene	2	Boulder Clay		Man
Tertiary	Pliocene	7		Alpine	
	Miocene	25			
	Oligocene	35			
	Eocene	55	London Clay		
	Paleocene	65			
Mesozoic	Cretaceous	135	Chalk		
	Jurassic	195	Oolitic Limestone		First Birds
	Triassic	225	Sandstone		First Mammals
Palaeozoic	Permian	280	Red Sandstone	Hercynian	
	Carboniferous	350	Coal, Millstone Grit, Limestone		
	Devonian	400	Old Red Sandstone	Caledonian	First Amphibians
	Silurian	425			First land plants
	Ordovician	500			First Fish
	Cambrian	600	N. Wales Slate		
PRE CAMBRIAN					Rare traces of very primitive marine life

oldest known rocks about 3 500 000 000 years.

Most sedimentary rocks contain fossils that enable geologists to build up a picture of life as it existed at the time of their formation. The picture changes as the rocks record the passage of the vastness of geological time. Fossils reveal the gradual evolution of life on earth, starting from the tiny marine organisms of Cambrian times, and culminating with the recent, the very recent, appearance of man himself.

2
The
Oceans

By far the greatest part of the surface of the earth is covered by water to varying depths. In fact, only a shade less than 30% of the earth's surface is high enough to protrude above the level of this water, the great bulk of which is contained in the basins of the Pacific, Atlantic and Indian Oceans. The Pacific alone covers over a third of the area of the earth's surface. For the most part, the limits of the oceans are well-defined, but occasionally they 'fray' at the edges and mingle with land areas to form marginal seas. The Caribbean, North and China seas are examples.

Relief of the ocean floor

Modern techniques of echo-sounding provide a precise and speedy method of determining the depth of the oceans. By plotting the soundings made by this method, an accurate and detailed contour map of the ocean floor may be built up. Such a map shows that the ocean floor is far from being the featureless plain that was once imagined. There are great contrasts in depth and the ocean floor has a relief that is little less diverse than that of the land surface. The main relief features of the floor of a typical ocean are indicated on the diagrammatic section included as Fig 17, and examples may be identified on the sketch map of the North Atlantic, Fig 18.

The *continental shelf* varies considerably in width. For the most part it is a narrow strip fringing the land areas, but locally it broadens out to form an extensive area of shallow water, everywhere less than 180 metres deep. Fig 18 shows that Britain rises from the shallow waters of the extensive continental shelf that fringes North-West Europe. In this Britain is fortunate for the world's richest fishing grounds are found in the shallow sunlit waters of the continental shelf where plankton is most abundant. The tapping of the gas resources of the rocks beneath the North Sea is greatly facilitated by the shallowness of the water in this area.

The continental shelf is of gentle relief, but towards its outer edge it is gashed by deep, steep-sided canyons which extend down the steep *continental slope,* the true edge of the continents, to the depths of the ocean beyond.

The *abyssal plains* occupy by far the greatest area of the ocean floor. They occur in well marked basins and are generally between 4500 metres and 5500 metres below the surface, but level plateau areas also occur at shallower depths. The relief is much more varied than their name would suggest. The monotony of the plains is frequently broken by extensive features that may be compared to mountain areas on land. Soundings also reveal the presence of numerous isolated hills or mountains known as *seamounts.* These are of volcanic origin and rise steeply from the ocean floor. The abyssal plains are generally blanketed by a layer of sediment, up to a kilometre thick, that has slowly accumulated over millions of years.

A *mid-ocean ridge* can be identified in all the oceans. In the Atlantic, as the Mid-Atlantic Ridge, it extends from Iceland in a great curving 'S', right to the Antarctic Ocean, where it is linked with similar features in other oceans. It is here that new oceanic crust is being formed as the Atlantic slowly becomes wider (page 8). Its highest parts, the summits of huge volcanoes, peep above the surface as small and lonely islands, such as those of the Azores group. The island of Tristan da Cunha, isolated in the South Atlantic, is another example. The volcanic origin of the Mid-Atlantic Ridge is emphasised by Fig 19. Here we see in progress the volcanic eruption that, in 1963, created the new island of Surtsey at the northern end of the ridge near the south-west coast of Iceland.

The oceans' greatest depths are recorded in very

Fig 17

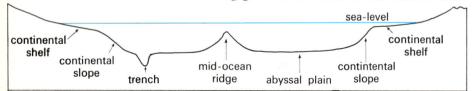

continental shelf — continental slope — trench — mid-ocean ridge — abyssal plain — contintental slope — continental shelf — sea-level

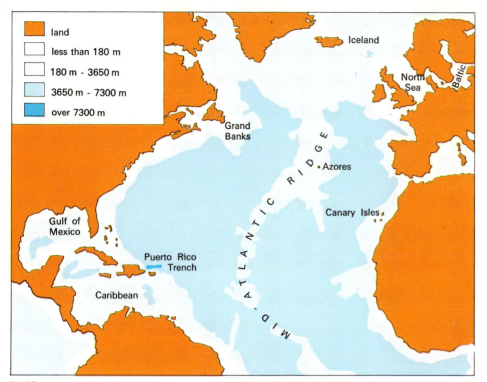

🟧	land
⬜	less than 180 m
⬜	180 m – 3650 m
🟦	3650 m – 7300 m
🟦	over 7300 m

Fig 18

Fig 19

restricted areas. They occur in long narrow trenches or *deeps* which all show depths greater than 7300 metres. These trenches, formed by the downward movement of plates, may be identified in each of the major oceans. The Puerto Rico Trench in the Caribbean is an example. They are most common in the western Pacific and in the Marianas Trench the record sounding of 11 035 metres was made. When it is realised that this depth exceeds the height of Everest by over 2100 metres, we can better appreciate the scale of the relief of the ocean floor.

Ocean currents

The waters of the ocean basins are in constant circulation. In detail, this circulation is very complex, but there are certain well-defined movements of surface water that are known as ocean currents. These currents seldom affect more than the top 180 metres of the oceans, and are generally slow moving. Speeds of about 0·5 km/h are typical, but in some cases the water drifts along at over 3 km/h.

Wind is the prime cause of ocean currents. Friction between air and ocean causes the surface water to be dragged along. The pattern of ocean currents is a reflection of that of the major winds over the earth's surface (Chapter 14) but the rotation of the earth and the shape of the land masses introduce modifications.

On average the surface temperatures of the oceans show a steady decrease from tropical to polar latitudes. The range is from over 25°C to a little above freezing point. Currents which transport warm water to colder parts of the oceans are known as warm currents. Movement in the opposite direction extends fingers of cold water into warmer areas and gives rise to cold currents.

Fig 20 shows the distribution of the important ocean currents. It shows that in each half of the major oceans there is a massive circulation, clockwise in the northern hemisphere and anti-clockwise in the southern, with a predominantly east-west movement about the Equator. Note particularly the currents in the northern Indian

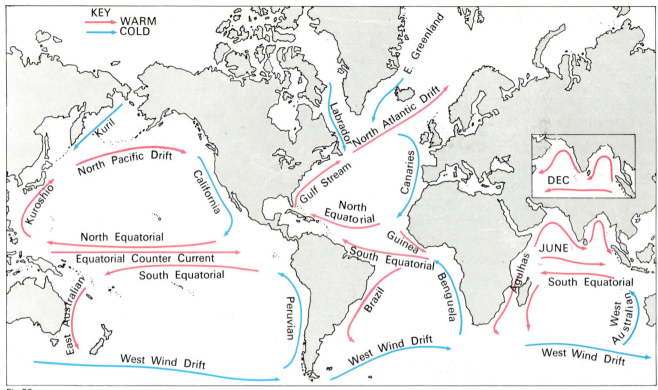

KEY
→ WARM
→ COLD

Kuril

North Pacific Drift

Kuroshio

California

North Equatorial

Equatorial Counter Current

South Equatorial

East Australian

West Wind Drift

Peruvian

Labrador

North Atlantic Drift

E. Greenland

Gulf Stream

Canaries

North Equatorial

Guinea

South Equatorial

Brazil

Benguela

West Wind Drift

DEC

JUNE

Agulhas

South Equatorial

West Australian

West Wind Drift

Fig 20

Ocean. They show a marked seasonal reversal in direction. This is the result of the monsoonal change of wind direction characteristic of the area, and underlines the importance of the wind as the main cause of ocean currents.

The chief significance of these currents lies in the influence they have on the climates of the land areas bordering the oceans. Examples are given in later chapters.

The salt of the sea

Examine Fig 21 which shows the *hydrological cycle*. Evaporation from the surface of the ocean produces water vapour which later condenses and falls as rain.

Of the rain that falls on the land some, as the diagram shows, is returned to the atmosphere by evaporation, and more returns through transpiration by plants. The remainder flows back into the oceans. On its journey over and through the rocks it collects a variety of minerals in solution. The amount of dissolved mineral matter in river water is extremely small, but these small amounts are constantly being added to the waters of the oceans. There the minerals accumulate, for they are left behind when sea water evaporates. Thus over the millions of years of geological time the proportion of dissolved minerals in sea water has been slowly increasing, until today they form, on average, 35 out of every 1000 parts of sea water. This *salinity* is usually expressed as $35\%_{oo}$.

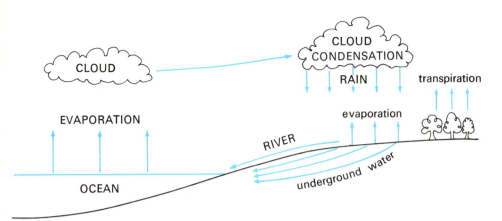

Fig 21

Calcium carbonate is the most common mineral carried in solution by rivers, but, as we have seen, it is removed from the ocean by marine organisms. This leaves sodium chloride (common salt) as the largest mineral constituent of salt water, accounting for over three-quarters of the total. Magnesium chloride and sodium sulphate are also present in appreciable quantities. The remaining 5% is made up of small quantities of nearly 50 elements, all in very low concentrations.

Common salt may readily be extracted from sea water by evaporation. In hot climates it is simply a question of allowing the sea to flow into shallow enclosed basins and letting the sun do its work. The valuable metal magnesium may also be obtained from the sea but, for this, elaborate technical equipment is necessary. The concentration of other minerals is too low to permit extraction on a commercial scale, but future technological development may enable man to tap the great mineral wealth at present 'locked up' in sea water. The humble seaweed could teach man a lesson in extraction. It is able to accumulate iodine to such an extent that it is an important source of that element.

Over the great extent of the open ocean, salinity shows only slight variations from the average figure of $35^{0}/_{00}$. It is slightly less than average in the polar oceans, but in tropical areas of high evaporation the figure rises to $37^{0}/_{00}$. Marginal seas show much greater variations in salinity. In the Baltic, for example, it is everywhere less than $10^{0}/_{00}$ and in places it is as low as $2^{0}/_{00}$. These low figures may be explained by the large additions of fresh water made by rivers and the low rate of evaporation in this cool area. In contrast, the Red Sea, which receives negligible contributions from rivers or rainfall, and where high temperatures promote much evaporation, salinity is of the order of $40^{0}/_{00}$.

It may be mentioned in passing that the saltiest salt water is not found in the sea, but in lakes that form part of areas of internal drainage. The Dead Sea is a good example. A high rate of evaporation causes a loss of water equal to that supplied by the river Jordan, and the Dead Sea has a salinity of $240^{0}/_{00}$. This high degree of concentration of minerals greatly facilitates their commercial extraction.

Sea-level?

That the level of the sea is not constant can readily be appreciated during a seaside holiday. The flowing tide, against which we are as powerless as King Canute, causes us to retreat up the beach, leaving our beach defences to be destroyed by the advancing water. Later the tide ebbs and we can regain our former territory.

There is a steady rhythm to the ebb and flow of the tide. Around our shores high and low water both occur twice in a period of just over 24 hours. In fact, the period between successive high tides is precisely 12 hours 26 minutes. If it is high tide at 10 a.m., the next high tide will be at 10.26 p.m. The next morning it will be at 10.52 a.m., and so on. The time of high tide is often chalked on notice boards at the entrance to the piers of the more popular holiday resorts, and with this information we can accurately predict the state of the tide for the rest of our holiday.

You have probably noticed that the tide does not come in the same distance each day. Many beaches show several lines of driftwood – tide marks – that are evidence of this. Similarly, the tide goes out to a different extent each day and so exposes a greater or lesser expanse of beach. This is because the tidal range, that is the difference in height between high and low water, shows daily variations. There is a regular rhythm to this, too, and it may be understood with the aid of Fig 22.

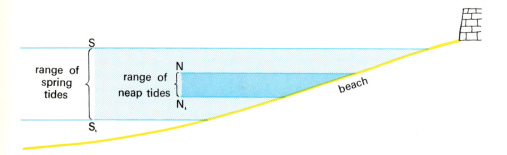

Fig 22

When the tide has its maximum range it is known as a *spring tide* and on the diagram the levels it attains are indicated by S and S_1. Over a period of about a week the range gradually decreases until it is at its minimum. This is a *neap tide* and is indicated by N and N_1. During the following week the tidal range shows a steady increase until the levels of a spring tide are reached again. This cycle is repeated so that spring and neap tides are recorded twice in each period of 28 days. The effect of the varying range of the tides on the extent of beach exposed may be gauged from the diagram.

It must be emphasised that the horizontal advance and retreat of the sea over a beach is purely an incidental result of the tides which are vertical movements of sea water.

The basic cause of tides is the gravitational pull of the moon, and, to a lesser extent, of the sun on the earth's water covering. These forces are constant over the earth, but due, among other things, to the differing depths and shapes of the ocean basins, tidal range shows great variations from place to place. Over the open ocean it is generally less than a metre, and it is even less in enclosed seas. The Mediterranean, for example, records tidal ranges of less than a third of a metre, and in the Black Sea the effect of tides is almost imperceptible. In the shallow waters of the continental shelf, the tidal effect is greatly accentuated. Around the coasts of the British Isles, although varying considerably, it is frequently more than 5 metres. It is greatest in shallow estuaries. In the Mersey estuary, the port of Liverpool experiences a range of spring tides as high as 10 metres, and even this high figure is exceeded in the funnel-shaped estuary of the Severn.

The great tidal range in the estuaries around the coasts of Britain has the advantage of allowing ports to be sited well inland in sheltered waters. However, large ships must await favourable tides before proceeding to and from the port and with a high tidal range expensive docks must be constructed to allow ships to load and discharge their cargo.

The highest tides in the world occur in the Bay of Fundy between the Canadian provinces of Nova Scotia and New Brunswick. Here the range of spring tides is 15 metres and fish are caught in nets suspended from poles firmly planted in the beach near low water mark. The 'fisherman' brings in his catch by horse and cart.

Knowing the levels to which the tide may rise, man has built dykes and other defences to protect low-lying land from its dangers. In instances that are fortunately rare, circumstances conspire to cause the sea to rise to levels higher than the highest normal tides. In the winter of 1953, strong northerly gales, coinciding with a spring tide, caused the piling up of water in the southern North Sea to a height of 3 metres above normal. This *storm surge* swamped the defences and flooded over large areas of the Netherlands and eastern England. It caused untold damage and the loss of over 2000 lives.

Fig 23

Fig 24

Submarine earthquakes occasionally set in motion huge waves which race across the ocean at speeds of hundreds of kilometres per hour. These waves are most common in the Pacific, and are known by the Japanese name of *tsunamis*. They gain in height in shallow water and crash against the coastline with disastrous effect. The tsunami that struck Japan in 1896 brought death to 27 122 people.

The average level of the sea is referred to as *mean sea-level* and is used as the basis for the measurement of altitude. But look closely at Fig 23, a view which was taken on the west coast of Scotland. Note the area of cultivated land sloping very gently seawards. Backed by steeply rising land it ends abruptly in a line of low cliffs just beyond the farmhouse. This feature is known as a *raised beach*, and a beach it was in former times. When mean sea-level was considerably higher than it is today, the steep slopes were high cliffs and the land that is today yielding good crops of grass and potatoes was washed by the tides. Other examples of this feature may readily be identified in Fig 24, another photograph taken of part of the west coast of Scotland. They occur in two sets at different heights, thus indicating more than one change of sea-level. Raised beaches are only one indication that mean sea-level has varied greatly in geological time.

Movements of the earth's crust have lifted rocks that were formed on the bed of the ocean to great heights above present sea-level: vast areas of land have been submerged to receive a new mantle of sedimentary rock. At any one time the amount of water imprisoned in the world's ice sheets and glaciers affects the amount of water in the ocean and so its level relative to the land surface. It is estimated that at the height of the Ice Age, sea-level was 100 metres lower than it is today, and that if all the world's ice were to melt, all land below the present 30-metre contour would be submerged.

Fluctuations of sea-level help to explain many features of present day scenery and their influence will be underlined in later chapters.

3
Unstable
Earth

Fig 25 shows, in broad outline, the major structural features of the earth's surface. The most impressive of these features, though by no means the most extensive in area, are the *new fold mountains*. These are long yet relatively narrow mountain systems of great complexity. They girdle the Pacific Ocean in a broken circle that includes the Rockies and Andes, each of which exceeds 6000 kilometres in length. Linked to this circle is a line of mountains that extends to the shores of the Atlantic and which includes the Himalayas, Alps and Atlas mountains among its better known ranges.

It is in the new fold mountains that the earth rises to its greatest heights. Peaks of over 4500 metres are commonplace and many rise to over 7000 metres, especially in the Himalayas where the mighty Everest towers to a record 8848 metres. The Himalayas provide a good example of a common feature of the new fold mountains, the occurrence of high plateaux between bounding ranges of greater altitude. The plateau of Tibet is the largest example, but others occur in the Andes, Iran and Turkey, and in western North America. In view of the great height of these mountains it is perhaps surprising to find that they

Fig 25

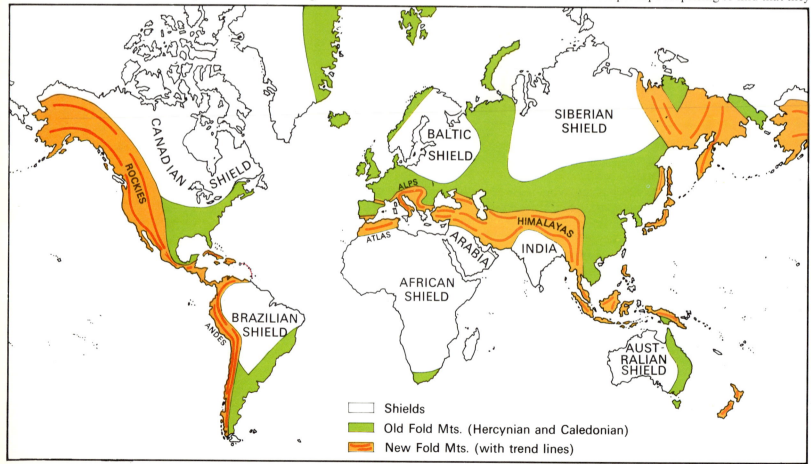

Shields

Old Fold Mts. (Hercynian and Caledonian)

New Fold Mts. (with trend lines)

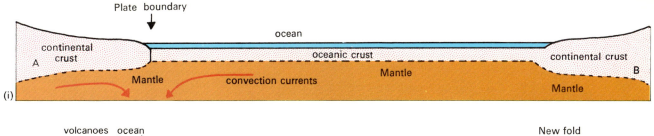

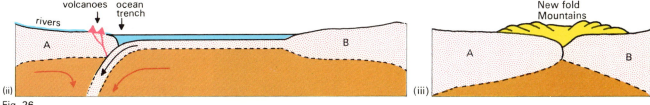

Fig 26

are mainly composed of sedimentary rocks that were formed on the ocean floor. These rocks have been contorted by intense folding and fractured by faulting on a large scale. Plutonic rocks occur in the very core of the new fold mountains, and volcanoes have added their contribution of volcanic rocks. In fact, volcanoes often sit on top of these mountain ranges, especially in the Andes, to form the highest peaks.

New fold (Alpine) mountains were created between 25 and 10 million years ago. Their impressive height is a reflection of their youth. They have been carved into the sharp and jagged outlines pictured in Fig 27, but time has been insufficient for the forces of denudation to bring about any great reductions in altitude.

The formation of fold mountains is undoubtedly a result of the interaction of mobile crustal plates, but the precise mechanisms involved are still not fully understood, and are the subject of active scientific research. Fig 26 outlines the current theory of mountain building in simple diagrammatic terms.

The boundary between two plates is indicated in Fig 26 (i). Convection currents within the mantle drag plate B

slowly but powerfully towards plate A. At the plate boundary, the oceanic crust is forced down into the mantle, which accepts and absorbs the incoming material (Fig ii). This downward movement has important consequences. The crust shudders with earthquakes and volcanic material escapes to the surface. In addition, a deep, narrow trench is formed in the ocean floor. Rivers erode the land surface, and the great thicknesses of sediment accumulate in the deep, still waters of the trench. These sediments may be as much as 10 km thick, and the lowest layers are converted to granite by radioactive heating. The ocean slowly shrinks as the two expanses of continental crust move ever closer to each other. Masses of relatively young rock are trapped in the jaws of a gigantic vice. Too light to be dragged down into the dense material of the mantle, they are eventually squeezed out onto the edges of the plates in huge and complex folds. Crustal fractures permit the addition of volcanic material. Should the downward drag of convection currents cease, the new-born mountains are uplifted to even greater heights. The final stage is represented by Fig 26 (iii). The ocean has vanished and two plates have been welded into one. Mountain ranges are the tombstones of departed oceans.

Fig 27

Fig 28

The Alpine Mountain Building Period has not been a unique occurrence. At earlier stages in the long history of the earth, similar periods have occurred in which mountains as high and majestic as the Alps have been created. Today these mountains, known because of their

antiquity as *old fold mountains*, show little of their former grandeur. They have, in the long eras since their formation, been worn down to modest altitudes and gentle outlines. They have been so buffeted by earth movements that often only the merest remnants remain.

The structure of the British Isles includes samples of these old mountain systems. The granite moorlands of Devon and Cornwall represent the plutonic heart of a mountain range that can be traced by similar remnants right across Europe and deep into Asiatic Russia. This range is known as the *Hercynian* and was formed some 300 million years ago.

The Highlands of Scotland are the remains of an even older mountain system, the *Caledonian*, which was at its highest over 400 million years ago. These mountains were reduced by denudation to a generally level surface of low altitude – a *peneplain*. Later they were uplifted and again attacked by agents of erosion which produced deep valleys. The peaks of the Highlands of Scotland are merely patches of the former surface left upstanding by erosion all around them. The contrast in scenery between new and old fold mountains can be appreciated by a comparison of Fig 27 and Fig 28. Note in the latter, taken in the Grampians, how the summits display a uniformity of height which is a clue to their formation by erosion of a former level surface.

The *shields* shown in Fig 25 are great areas of the earth's crust that have remained relatively undisturbed for much of geological time. Faint traces remain of mountain ranges long since worn away. Shields are mainly composed of ancient plutonic and metamorphic rocks that are frequently the source of rich mineral deposits. These shields have been affected differently by geological processes. That of South America, for instance, has been tilted. Parts of it are concealed by layers of sedimentary rock which form wide lowland plains. The shield which forms the great bulk of the continent of Africa remains a plateau of considerable height. From this plateau, rivers

Fig 29

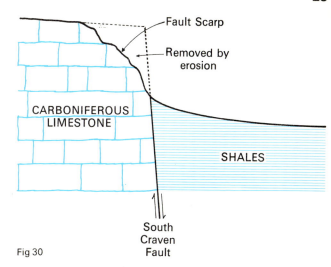

Fig 30

descend to the ocean in a series of waterfalls and rapids. These hamper navigation but offer a tremendous potential of hydro-electric power that has yet to be harnessed. The shields of Canada and Scandinavia show evidence of the erosive power of ice in such a multitude of lakes of all shapes and sizes that large areas of the surface are covered more by water than by land.

Features due to faulting

Vertical movements of the earth's crust along a fault frequently leave their mark on the landscape. Fig 29 shows a simple example, the South Craven Fault, near the village of Giggleswick in West Yorkshire. The low land to the right of the road in the photograph has slipped down along a fault, leaving the limestone to the left upstanding as a steep and imposing slope known as a *fault scarp*. The situation is illustrated in Fig 30. Not every fault shows a fault scarp because denudation so smooths the surface that the presence of a fault can usually be detected only by the arrangement of the rock, as in Fig 13.

When faults occur in parallel lines, they often give rise to distinctive features of the landscape. Study Fig 31. In the centre of the photograph the River Jordan flows over an area of level land to join the Sea of Galilee. In the distance, the straight valley side rises steeply for over 600 metres. The slope in the foreground is equally steep. This is a *rift valley*, and Fig 32 shows that the area of the photograph is only a tiny part of the huge trough in the earth's crust that can be traced from the borders of

Fig 31

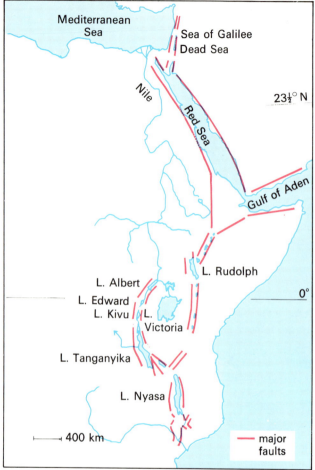

Fig 32

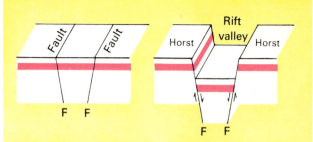

Fig 33

Turkey, through the continent of Africa, almost to the south Indian Ocean. Very narrow in relation to its immense length, it is bounded by parallel faults. Between the faults, rocks are found to occur at lower altitudes than they do in the bounding uplands which are known as *horsts* or *block mountains*.

Plate movements are responsible for the formation of major rift valleys. The Red Sea, for instance, is the result of the Arabian Plate (Fig 4, page 8) pulling away from that of Africa, and is an ocean in its infancy. Crustal tensions caused by this movement have acted on lines of weakness to create the complex rifts of east Africa. A rift valley, the Thingvellir Graben, crosses Iceland, which is slowly becoming wider as the American and Eurasian plates drift apart. Small rift valleys may be regarded as minor adjustments to stresses and strains within the crust. The River Rhine between Basle and Bingen flows along the length of the Rhine rift valley. The Central Valley of Scotland is another example. The straight lines of its boundary faults can be easily identified on the geological map of Great Britain in your atlas.

When the land between parallel faults rises rather than subsides, the resultant feature is a horst. Fig 34 explains their formation. The Harz mountains of Germany are a good example.

Fig 34

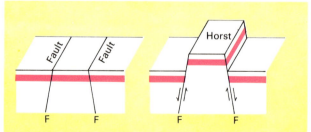

Vulcanicity

Volcanic activity is perhaps the most sudden and spectacular of all geological processes. It is a process

Fig 35

Fig 35 shows the island volcano of Surtsey in the course of eruption. The large cloud, reddened by the glowing lava, seen rising above the crater, is composed of the gases which are a feature of all volcanic outbursts. It is for the most part water vapour, but will contain other gases such as carbon and sulphur dioxide. These gases normally disperse quietly and harmlessly into the atmosphere, but they are not without their dangers. In 1902, for example, a dense cloud of suffocating gas, charged with tiny particles of hot lava, rolled down the sides of Mt. Pelée, a volcano in the West Indies, and wiped out the town of St. Pierre and its population of nearly 30000 people. Occasionally the water vapour condenses into torrents of rain of such intensity that they may prove to be the most destructive part of the eruption.

Look again at Fig 35. Red hot lava with a temperature of perhaps 1000°C spills over the edge of the crater and flows down to the sea as a fiery river of basalt. Soon it will cool into solid rock and so increase the size of the volcano. Lavas vary in their composition. Some, known as *acid* lavas, are viscous and do not flow far before solidifying and so build up steep-sided *lava cones* (Fig 36 (i)). At the other extreme, *basic* lavas will flow great distances and build up lava cones with very gently sloping sides (Fig 36 (ii)). The volcanoes which make up the island of Hawaii are of the latter type. The highest is called Mauna Loa and it rises from the ocean floor to an altitude of 4170 metres above sea-level.

Fig 36

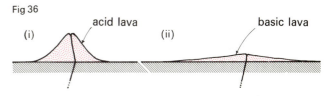

mainly confined to the margins of the earth's crustal plates. Where plates are drifting apart, molten material rises to the surface from the hot layers within the mantle. Normally it spreads out to form new ocean floor, but locally it may be piled up into huge volcanoes, such as those which form the island peaks of the Mid-Atlantic Ridge (page 14).

Active volcanoes are commonly found where one plate is sliding beneath another. As the oceanic crust descends into the interior, its temperature rises and gases are liberated on a large scale. This produces a mobile mixture of gas and liquid rock, which rises to the surface along any convenient line of weakness in the overlying continental crust.

The broken mountain rim of the Pacific Ocean is the scene of frequent demonstrations of the dramatic power of vulcanicity. Japan alone has over twenty *active* volcanoes which have erupted in recorded history. This number is only a fraction of the total number that dot the landscape of this mountainous country. Many of them are *extinct* and really belong to former geological periods, but for many others, classed as *dormant,* the possibility of future activity cannot be ruled out.

Volcanic eruptions need not be restricted to a single point of eruption. In some instances, not, fortunately, in recent times, basic lava has welled up along the whole length of a series of faults and spread widely over the earth's crust. Successive eruptions build up the land

Fig 37

surface to the dimensions of a plateau. The Deccan region of India is an example of one of these areas of *plateau basalts* as they are known. Northern Ireland provides us with another example. Look closely at Fig 37 which is a sunshine and shadow view of Antrim's Pleaskin Head. Several distinct layers of rock may be identified in the cliff face. They are of varying thickness and show the characteristic columnar jointing of basalt.

Fig 38

Each layer is the lava from one of the many eruptions this area experienced in the distant past.

Fig 38 is of Fujiyama the volcano which rises gracefully to a height of 3776 metres just west of Tokyo, the capital city of Japan. The lower, more gentle slopes are marked by a pattern of small fields and are intensively cultivated. The Japanese farmer is attracted to these slopes by the great fertility of their volcanic soils. This fertility more than compensates for the danger of living on a volcano. Fujiyama is an example of a *composite* volcano. It has been built up not of lava alone, but of alternate layers of lava and the solid products of eruption such as volcanic ash known collectively as *tephra*.

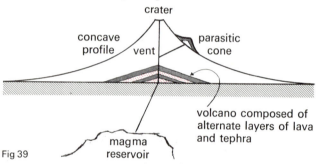

Fig 39

Chimborazo and Cotopaxi in the Andes, Vesuvius and Etna in Italy and, indeed, the majority of the higher and better known of the world's volcanoes are of this type, which is illustrated by Fig 39. Following an eruption of lava, the vent becomes sealed as the lava in the vent solidifies, and the way is only reopened when the pressure of rising gases is powerful enough to explode away the blockage. Such explosions create tephra. Solidified lava from former eruptions is reduced by the force of the explosion to particles of solid material ranging in size from large blocks to the finest dust.

Fig 40 shows tephra being formed in another way. Askja, an Icelandic volcano, is in eruption and a fountain of liquid lava spurts high into the air. Particles of lava cool and solidify in mid-air and fall to earth in the dark cloud of solid material that we see in the photograph. Tephra

Fig 40

tephra. Its excavation has given us a clear picture of urban life in Roman times. In 1883 the volcanic island of Krakatoa, near Java, was destroyed by what must have been the greatest explosion the world has ever heard in historic times. Three tiny islands were all that remained, but since then a fourth has appeared, the summit of a submarine volcano appropriately named Anak Krakatoa, or 'Child of Krakatoa'.

Although sometimes giving warning of their intentions in a series of earthquakes, volcanoes are generally unpredictable. In 1961 the population of Tristan da Cunha was driven from its lonely island home by unexpected outpourings of lava from the volcano that was thought to be extinct.

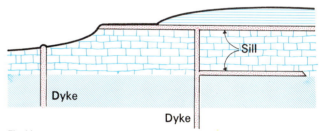

Fig 41

blankets the products of former eruptions and is, in its turn, buried by subsequent lava flows. Thus, layer upon layer, the volcano gains in height as eruption follows eruption. Sometimes, the rising lava finds an easier route to the surface and breaks out on the side of the volcano to form a *parasitic cone*. The slopes of Etna are marked by over 200 of these mini-cones.

Volcanoes show great individuality: there are almost as many types of eruption as there are volcanoes. Stromboli, off the coast of Sicily, erupts gently at least once an hour on average. Others have more violent eruptions at irregular intervals. In A.D.79, Mt. Vesuvius exploded with such force that the city of Pompeii was buried in

Volcanic activity does not always extend to the surface of the earth. Magma may be intruded into the crust, there to solidify and to be revealed only when erosion has stripped off the overlying rock. We have seen, in Chapter 1, how large masses of plutonic rock are placed in the heart of fold mountains. Two other ways in which magma is intruded are illustrated in Fig 41. Magma which solidifies in a vertical crack in the rock is known as a *dyke*. An intrusion which follows the bedding planes is called a *sill*. It was on the top of the steep scarp slope formed by an outcrop of Whin Sill that Hadrian built his famous wall across northern England.

4
Weathering

Fig 42 shows part of the fabric of St. Mark's Church, Bristol. The decorative carving, done with loving care by skilled masons in the 13th century, is obviously not in its original condition. Its decay is due simply to its exposure to the atmosphere over the last 750 years or so. The same decay can be observed in the humble bricks and mortar of the average house. The mortar, being the weaker, suffers the most and houses have to be 'pointed' occasionally to make good the damage. Any rock exposed to the atmosphere will eventually be reduced to small particles and this process is known as *weathering*.

Fig 43 shows an isolated boulder exposed in a small Yorkshire valley. The atmosphere has in its armoury several slow but powerful weapons for its irresistible attack on this and all rocks. When, for instance, rock is exposed to the sun, its surface is heated and expands, only to contract when the temperature falls. Expansion and contraction frequently repeated give rise to stresses and strains sufficient to cause cracks in the surface of the rock. In winter, rainwater collects in these and other cracks and crevices and frequently turns to ice. As water freezes, it increases in volume and the force of this

Fig 42

Fig 43

Fig 44

combined attack of these various processes, rock is gradually broken down into small particles.

On steep slopes the products of weathering fall under the influence of gravity, and accumulate at the foot of the slope. In mountainous districts, the peaks are often shattered by freeze-thaw action and sharp angular fragments of rock accumulate as *scree* at the sides of the valley below. An example may be found on the right-hand side of Fig 107 (page 53).

The finer particles of rock produced by weathering form the basis of soil. On slopes the roots of growing vegetation tend to bind it in place, but even on a well-grassed slope the material is gradually moving downhill, lubricated by water in the soil. This movement, like most geological processes, is so slow that it is imperceptible, but close observation reveals telltale signs that it is taking place. Look at the valley side in Fig 44. Soil moving down the valley side piles up on the far side of the wall causing it to bulge dangerously in places. Soil slipping away from the tree in the centre of the photograph has left the roots exposed. The tiny terraces on the valley side are further evidence that the soil is moving downhill, eventually to be carried away by the stream in the foreground. This downhill movement of weathered material is known as *mass wasting*. It plays a significant part in the shaping of the landscape.

It is unwise to remove the vegetation from sloping land, for without its protective cover, the soil is soon washed away. In many parts of the world farmers have cultivated such land only to suffer the serious consequences of *soil erosion*.

expansion is, in time, sufficient to split the hardest rock. This very effective process is known as *freeze-thaw*.

The boulder in the photograph is composed of limestone which is slightly soluble in rainwater. Other chemical processes, speeded up by the liberation of acids from decaying vegetation, contribute to the break-up of the rock. Even the tree growing in a pocket of wind-borne soil has its part to play. The growth of its roots subjects the rock to an additional stress. As a result of the

5

The Work of Rivers

Running water is one of the powerful forces which carve the rocks of the earth's crust into the finely detailed scenery we see about us. It operates slowly but persistently and, given sufficient time, can create canyons nearly 2 kilometres deep, or reduce mountain areas to the dimensions of a low lying plain. Rivers, like all agents of erosion, have three tasks to perform. They erode or wear away the rocks of the earth's surface and then transport the debris which is later put down or deposited.

A river may erode in a variety of ways. Running water alone may wear away the rock, but its erosive power is vastly increased if, as is usual, it carries a load of rock particles. These rock particles are swept downstream by the current, and bump and bang, scratch and scrape against the bed of the stream gradually wearing it away. This downcutting is particularly effective when a pebble caught in a slight hollow in the bed of a stream is whirled around by the force of the current. It acts like a drill, boring deeply into the rock. The pebble is worn down and washed away, only to be replaced by another, and so the work goes on. The rock in Fig 45, although only submerged in time of flood, is pitted with *pot-holes* produced in this way. The rock in the foreground of Fig 46 has been carved and fluted by the same process.

Fig 45

Fig 46

In limestone areas, another factor helps in the lowering of the bed of the stream, for this rock, being soluble, is slowly removed in solution.

The effectiveness of erosion depends on the nature of the rock over which the stream flows. Soft, weak rocks are worn away more readily than those which are tough and resistant. Under normal conditions erosion is a very slow process. It is in time of flood that most of the damage is done. At these times, when the volume and speed of the smallest stream may be dramatically increased, even large boulders can be sent crashing down the valley, and erosion is more rapid.

The load of a river, which consists of fragments of rock derived from the weathering of the valley sides as well as by the river's erosive action, is transported downstream. On the journey, these rock fragments are gradually reduced to a small size by contact with each other and the bed of the stream. The amount of load a river can carry depends on its speed and volume. If a river is carrying its maximum load, and its speed is reduced, then part of the load must be put down or deposited. Deposition is extensive in the lower stretches of the river valley and leads, as we shall see, to the formation of important features of the landscape.

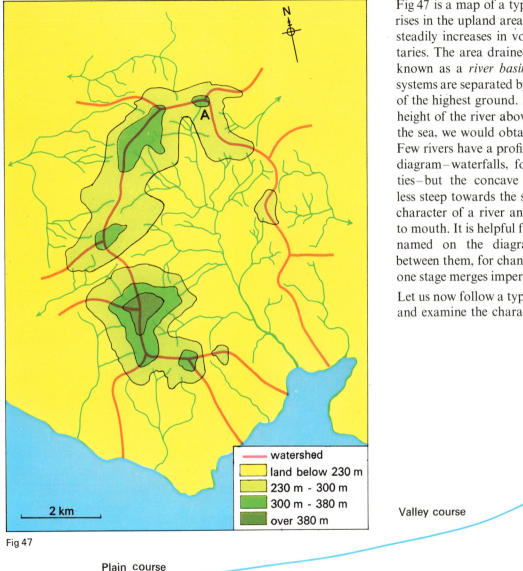

Fig 47

Fig 47 is a map of a typical river system. The main river rises in the upland area at A. On its journey to the sea it steadily increases in volume as it is joined by its tributaries. The area drained by a river and its tributaries is known as a *river basin*. The basins of adjoining river systems are separated by *watersheds* which follow the line of the highest ground. If we were to plot on a graph the height of the river above sea-level against distance from the sea, we would obtain a curve such as that in Fig 48. Few rivers have a profile as smooth as that shown in the diagram—waterfalls, for instance, introduce irregularities—but the concave profile, becoming progressively less steep towards the sea, is typical of most rivers. The character of a river and its valley changes from source to mouth. It is helpful for us to recognise the three stages named on the diagram. No divisions are marked between them, for changes take place very gradually and one stage merges imperceptibly into the next.

Let us now follow a typical river on its journey to the sea and examine the characteristic features of each stage.

watershed
land below 230 m
230 m - 300 m
300 m - 380 m
over 380 m

2 km

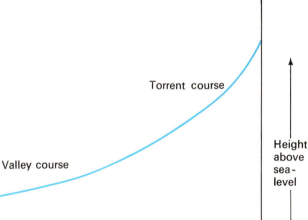

Fig 48

Torrent course

Valley course

Plain course

Sea-level

Sea

Distance from sea

Height above sea-level

Fig 49

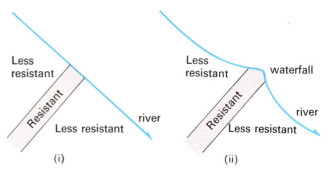

Fig 51

The river flows down

Look first at Fig 49. This shows a river near its source in its torrent stage. The valley sides are steep and the small stream occupies the whole of the valley floor, giving a steep V-shaped cross-section (Fig 50). The river, which is actively eroding its bed, flows round a series of interlocking spurs and its course is marked by many small waterfalls and rapids.

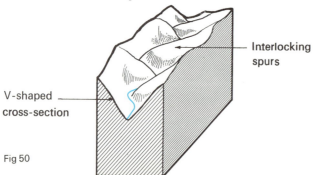

V-shaped cross-section

Interlocking spurs

Fig 50

Waterfalls are a common feature of the torrent stage. They may be formed in a variety of ways but the great majority are created by the erosive work of the river itself. Fig 51 (i) illustrates a common situation. The rocks over which the river flows include a band of resistant rock. This is eroded less than the weaker rocks above and below and so, in time, the profile becomes that indicated in Fig 51 (ii). Now there is a definite break of slope that will be marked by a waterfall or rapids.

An interesting type of waterfall occurs where the resistant band of rock is horizontal. Thornton Force in north-west Yorkshire is a good example, and is the subject of Fig 52. The resistant strata is limestone which lies above less resistant slates. The river tumbles over the edge of the limestone into the *plunge pool* in the foreground. It may be possible to identify two figures standing behind the curtain of falling water and this indicates that the rock there has been hollowed out. Fig 53 shows the situation in cross-section. Spray from the waterfall wears away the weaker rock, and this undercutting causes the edge of the limestone to collapse. As this process is repeated the waterfall retreats upstream, leaving a gorge behind it as it does so. The large block of rock in the foreground of the photograph is a remnant of a former collapse.

Fig 52

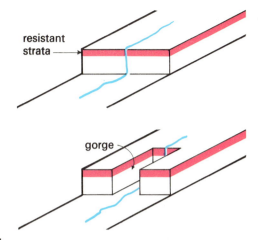

Fig 54

The famous waterfall at Niagara, on the border between the United States and Canada, is of this type. It is eating headwards at a rate of a metre a year, and this retreat has created a deep gorge over eleven kilometres in length, Fig 54. Waterfalls as high and as powerful as Niagara offer a tremendous source of hydro-electric power.

The cross-section of a valley depends upon the balance between river erosion and mass wasting. As Fig 55 shows, downcutting by the river is only responsible for excavating a small part of its valley. It is mass wasting (Chapter 4) which produces the characteristic V shape. If, for some reason, mass wasting is not active, then the river flows in a narrow, steep-sided trench known as a *gorge*.

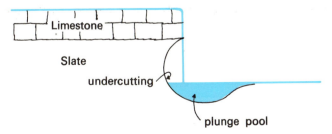

Fig 53

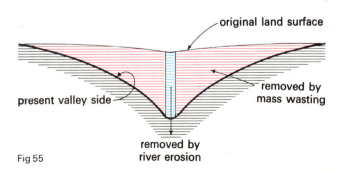

Fig 55

Fig 56

Fig 57

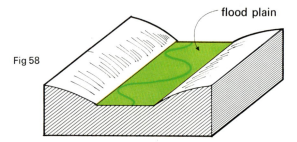

flood plain

Fig 58

Fig 59

The gorge of the Colorado, the Grand Canyon, shown in Fig 56, is the best known example of this feature. Here, in this desert area, there is little moisture to encourage mass wasting, and the river has cut down to a depth of 1905 metres.

Fig 57 shows a river in its valley stage, and we can note the changes that have taken place. Gone are the interlocking spurs. The sides are less steep and further apart, and the river *meanders* over a level valley floor. Its cross-section may now be represented by Fig 58.

Fig 59 will help us to understand how these changes have been brought about. The current is swiftest and erosion most powerful on the outside of the bends in the river. Here, the river is constantly nibbling away at the bank, undercutting it and causing it to slump into the water. The river bank on the outside of the bends is gradually being pushed further and further back. The river channel is not, however, getting any wider for on the inside of the bend, where the river is slow-moving, land is being created by deposition, Fig 60.

Fig 60

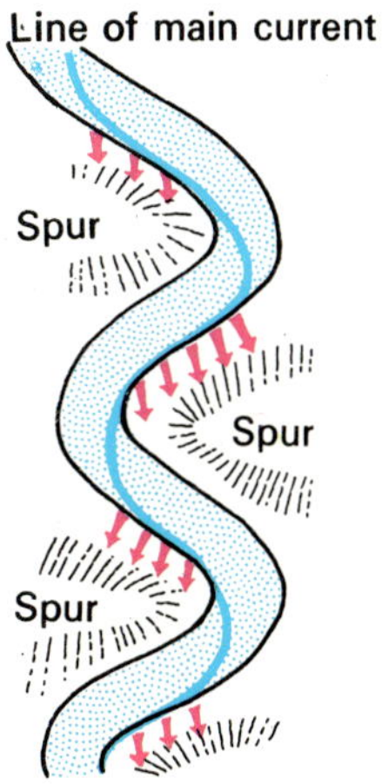

Now look at Fig 61. The main current hugs the outside of the bends and erodes the valley sides in the direction of the arrows. As a result of this *lateral erosion,* spurs are removed and the valley sides are pushed back thus creating a level floor to the valley. This is known as the *flood plain* and it receives layers of fine sediment when the river overflows its banks.

The process of valley widening continues with increasing effect as the river journeys on to the sea. In the plain stage, lateral erosion becomes the main outlet for the river's energy, for downcutting has ceased. Here the river meanders widely and freely over an extensive flood plain. The valley sides are low and far apart and the cross-section is shown in Fig 62.

Line of main current

Spur

Spur

Spur

Fig 61

Fig 62

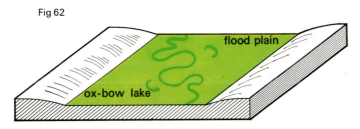

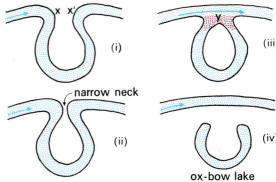

Fig 63

Flood plains, especially in the lower reaches of large rivers, are often scarred by small, curved, narrow lakes which look like detached portions of the river. Appearance is not deceptive, for that is just what they are. They are called *ox-bow lakes* and Fig 63 explains their formation. Fig 63 (i) shows a typical meander. The main current, eroding on the outside of the bends at X and X^1 brings the two arms of the meander closer together until they are separated only by a narrow neck of land. Eventually the river breaks through this neck and, taking the easiest and steepest course, bypasses the meander which becomes a stagnant backwater. We see a river at this stage in Fig 64. Deposition takes place at Y, Fig 63 (iii), and so the meander is sealed off from the river and becomes a lake. Ox-bows are soon filled by deposition from flood waters.

Fig 64

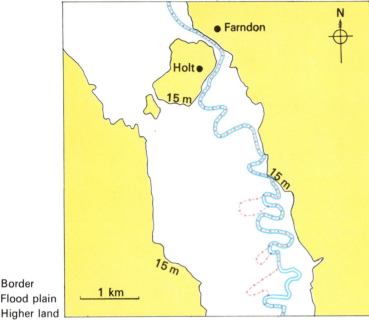

Border
Flood plain
Higher land

Fig 65

The creation of an ox-bow lake represents a change in the course of the river, and as a result of this process, a major river is constantly shuffling across its flood plain. These changes of course are, in geological terms, both frequent and rapid. Fig 65 shows part of the flood plain of the River Dee a few kilometres south of Chester. The channel of the Dee was chosen as the border between England (on the east) and Wales. Today, as the map shows, river and border do not coincide. The latter follows the course of the river as it was centuries ago.

To reach a really large river such as the Mississippi one has, surprisingly, to travel uphill. The river flows above the level of the surrounding land and the highest parts of the flood plain are the *levees* which border the river. Fig 66 illustrates the situation. The water in contact with the bed of the river is slowed down by friction and deposition takes place. Thus the river channel is gradually built up. The water that escapes over the banks of the river in

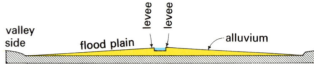

Fig 66

time of flood also suffers a reduction in speed. The deposition that results is greatest alongside the channel and becomes less and less the further the floodwater flows. Hence the levees and gently sloping flood plains of major rivers.

When a river flows above the level of the surrounding land, there is an obvious risk of serious floods. In spite of this danger the flood plains of major rivers are often closely settled by agricultural populations seeking the benefits of the high fertility of alluvial soil. Floods on mighty rivers such as China's Hwang Ho can have disastrous results and cause terrible loss of life. Man defends himself by building strong embankments along the river to keep it in its place. This may work for a while but such embankments cause the river to raise its bed more quickly and the risk of future flooding is increased. A better defence is to imitate nature by cutting artificial channels through the necks of the meanders. This shortens the course of the river and increases its rate of flow.

When, at the end of its journey, the river meets the sea, its speed is abruptly checked and extensive deposition takes place. If more material is deposited than can be removed by the sea, sediment slowly accumulates. New land is created as a flat, low lying plain is built out into the sea. Such a feature is known as a *delta*. On the flat delta surface, the river splits up into many *distributaries* which follow their own meandering paths to the sea. Changes of course are frequent and the land is scarred by old stream channels and shallow lakes.

Deltas are most common where large rivers flow into quiet seas. The Mediterranean, for instance, contains many examples, among them the Nile, Po, Tiber and Rhone. The densely-populated delta of the Ganges–

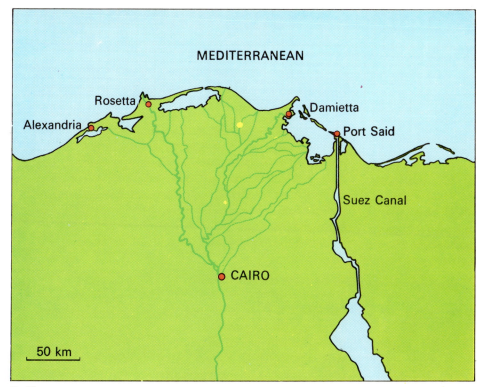

Fig 67

Fig 68

rate of 60 metres a year. The still waters of lakes favour delta formation, such as the large example built up in the Caspian Sea by the River Volga. Deltas vary greatly in outline, but the classic shape is that of the Nile, Fig 67. Indeed, the name 'delta' is derived from the resemblance of this area to the fourth letter of the Greek alphabet.

Compared with the mighty examples mentioned above, that of Scotland's River Drynoch is no bigger than a postage stamp. Yet, as Fig 68 shows, it well illustrates the characteristic features of the typical delta. Fig 69 shows a delta extending into the clear, still waters of a Norwegian fjord. The stream of muddy water in the right foreground is evidence that the work of delta formation is still in progress.

Fig 69

River terraces

Look again at Fig 48 which shows the profile of a typical river. Downward erosion is controlled by mean sea-level. In the plain stage when the height of the land is close to that level, downcutting ceases. But, as we saw in Chapter 2, mean sea-level is not constant and is subject to fluctuation. Imagine the situation if the sea-level should fall. The river must flow down a steeper slope to reach the sea, Fig 70. It will have greater speed and downcutting will be renewed. The river is youthful again: it has been

Brahmaputra occupies an area greater than the United Kingdom and that of the Mississippi is growing at the

rejuvenated. The revitalised river cuts downwards as far as it can and then shuttles back and forth, eating into and removing the flood plain that it so laboriously built up.

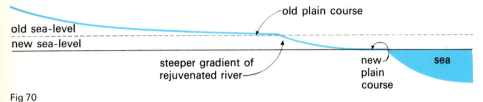

Fig 70

Look closely at Fig 71. The football pitch in the foreground is on the level flood plain of the River Mersey. Beyond the far goal-post the land rises steeply to a higher level, and the houses in the background are on a higher level still. Thus the land rises in terraces above the flood plain. These are known as *rejuvenation terraces*, and they occur in pairs at the same altitude, one on each side of the river.

Fig 71

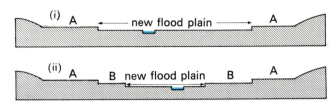

Fig 72

Fig 72 (i) shows the new cross-section. Remnants of the old flood plain remain at A as terraces standing above the new flood plain that the river is busily creating. The river, given sufficient time, may remove all its old flood plain but it is usually the case that some patches remain. A second rejuvenation sees the process repeated. Down-cutting starts again and then lateral erosion carves the second flood plain into a second set of terraces, Fig 72 (ii).

In the valleys of many British rivers three pairs of rejuvenation terraces may be identified. The lower Thames, for instance, has terraces at approximately 30 m, 15 m, and 4 m above the present flood plain. Terraces often provide dry settlement sites in an otherwise ill-drained area, and Britain's larger rivers are often flanked by lines of villages perched on the highest terrace. Fig 73 shows the pattern of terraces of the Mersey, west of Stockport.

Fig 73

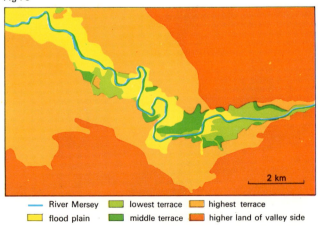

| — River Mersey | ▢ lowest terrace | ▢ highest terrace |
| ▢ flood plain | ▢ middle terrace | ▢ higher land of valley side |

Drainage basins

Although river basins show great variety in the arrangement of their member streams, certain common patterns may be recognised. One is shown in Fig 47. This pattern, where the tributaries are to the main stream as branches to a tree, is described as *dendritic,* and Fig 74 shows an example on a Lake District hillside. When the land is in the form of a dome, drainage takes on a *radial* pattern

Fig 74

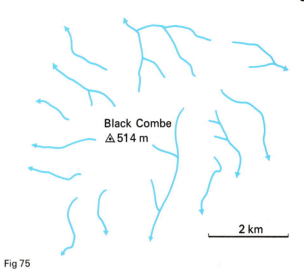

Fig 75

(Fig 75). *Trellised* drainage, where tributaries are arranged more or less at right angles, is illustrated in Fig 76.

Fig 76

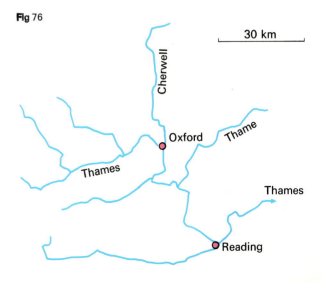

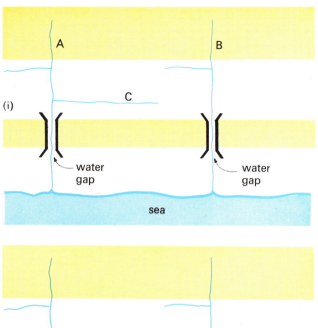

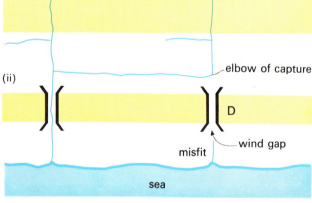

Fig 77   resistant strata ☐

drainage developed on alternately resistant and less resistant strata. It is a situation common among the scarplands of South-East England. The main rivers, A and B, cut through the resistant strata by means of well marked *water gaps*. A is assumed to be the more powerful river and its tributary, C, eats headwards along the outcrop of weaker rock. It does this so successfully that when it reaches river B it is at a lower level. River B is diverted into C and so into A. River B has been captured.

The new pattern of drainage is shown in Fig 77 (ii). The point of capture is marked by a sharp bend known as the *elbow of capture*. The beheaded remnant of river B is now a tiny stream in a large valley – a *misfit*. What was a water gap at D is now without a river and is known as a *wind gap*. The increased power of river A may now encourage it to capture other adjoining streams and so extend its drainage basin still further.

Fig 78

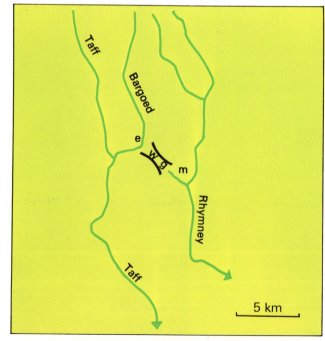

The boundaries between drainage basins are not permanent. By headward erosion, streams are gradually lengthened as their sources are pushed further and further inland. This causes a migration of watersheds and changes in the relative importance of adjoining river basins. Sometimes, by a process known as *river capture*, a river may greatly extend, in a most unfriendly fashion, the area under its control at the expense of a neighbour. Study Fig 77 (i). It shows a typical area of trellised

Fig 78 shows a situation where river capture has taken place in South Wales. A tributary of the Taff has eroded headwards to capture the Bargoed, which formerly joined the Rhymney. We can readily identify the telltale signs of elbow of capture (e), wind gap (wg) and misfit stream (m). It illustrates the point that river capture is not restricted to areas of trellised drainage. Your atlas will provide you with an example on a larger scale. Turn to the map that shows the rivers of Northumberland. The North Tyne and its extension, the Rede, have eroded northwards along an outcrop of weak rock and in so doing have captured the headwaters of several streams, the beheaded remnants of which flow eastwards to join the North Sea.

To work out the development of the drainage sketched in Fig 79 will prove a rewarding exercise.

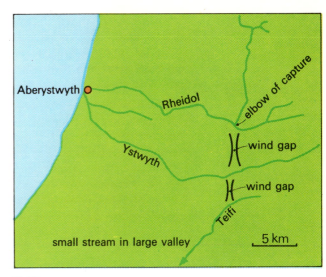

Fig 79

6
Water
Underground

As we saw in Fig 21, not all the rain that falls on the land collects into rivers and journeys to the sea over the surface of the earth. Some is returned to the atmosphere and more sinks into the rocks to become the *ground* or *underground water* we are concerned with in this chapter.

Not every type of rock permits the downward movement of water. Clay, for instance, is made up of particles so small and so closely packed that water is unable to pass through it, and this rock is described as *impermeable*. On the other hand, sandstone and chalk are *permeable* because, between their constituent particles, there are tiny spaces which allow the slow passage of water. Water may also penetrate a rock by means of joints and faults. Indeed, some rocks such as granite, which are normally impermeable, may contain large quantities of underground water if they are particularly well-jointed.

The easiest way of tapping these supplies of underground water is to dig a well. A well is simply a hole in a permeable rock deep enough to reach water. A diagram of one is given in Fig 80. Below a certain level the rock is saturated with water which seeps into the bottom of the well. The level of water in the well corresponds to the upper level of saturated rock, but this is not constant. It rises in periods of heavy rain and falls in time of drought. In serious drought it may fall below the bottom of the well, which then becomes dry.

Fig 80

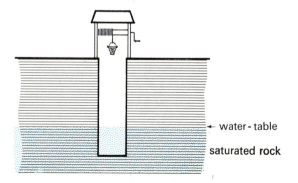

← water-table

saturated rock

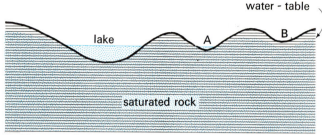

Fig 81

The upper level of saturated rock is known as the *water-table*. This, perhaps, is not the happiest of terms, for it is far from level. It arches up under hills and slopes down into valleys. Fig 81 illustrates the characteristic ups and downs of a typical water-table. Note how closely it is related to water features on the surface – it corresponds in height to the surface of rivers and lakes. The water-table, remember, is not constant. A slight rise and a river would appear in the valley at B. A movement in the opposite direction would cause the river at A to run dry. As the water-table is nearest the surface in low-lying areas, that is where wells are best sited. When Jack and Jill went up the hill to fetch a pail of water, it was not a well that they were visiting, but a hillside spring.

Fig 82

A *spring* is the name given to any natural outflow of water from the rocks. They are commonly found where permeable rocks lie above an impermeable layer. Fig 83 is a diagram of the spring shown in the photograph on page 42. The well-jointed limestone is saturated to the level of the water-table and water, guided by the joints, flows out at the junction with the impermeable rock and runs down the hillside as a tiny stream. Many large rivers originate from such small beginnings.

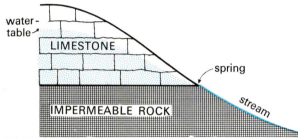

Fig 83

Springs commonly occur at the foot of the steep, scarp slopes of chalk cuestas and Fig 84 explains why. Large villages have grown up around these powerful springs and the line of villages sited at regular intervals along the scarp foot is known as *spring-line settlement*.

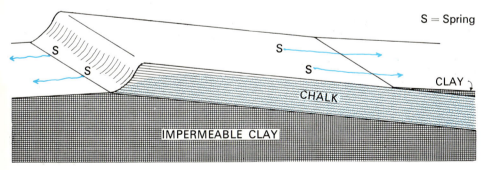

Fig 84

Chalk is a highly permeable rock. The rain it receives sinks underground leaving the surface virtually devoid of water. In times past, conditions must have been different, for chalklands are marked by complex valley systems which, although undoubtedly eroded by water, are now

Fig 85

completely dry. Fig 85 shows a typical example from the Yorkshire Wolds. Where streams do occur on chalk they are small and are found only in the deepest valleys. Many of them only flow when, after periods of heavy rain, the water-table is unusually high. These intermittent streams are known as *bournes*.

The Chiltern Hills and the North Downs are chalk cuestas, and Fig 86 shows that they are parts of the same great syncline. Note that the water-table in the uplands is higher than the surface of the London clay. This means that the water in the rock is under considerable pressure. When a well is driven through the clay and into the chalk below, this pressure is sufficient to drive the water to the surface or even above it as a natural fountain. Such a well is known as an *artesian well*. In London, however, so many of these wells have been bored to provide water for domestic and industrial purposes that the water-table, and hence the pressure, has fallen, and water has now to be helped to the surface by pumps.

Fig 86

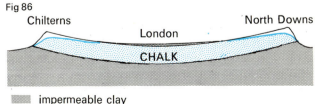

impermeable clay

Fig 87

Supplies of underground water are of special significance in arid areas of low and unreliable rainfall. Water from Australia's large artesian basins has made possible the extensive grazing of sheep and cattle over much of the dry interior of the continent. The development of the mineral resources of the Sahara desert has been greatly facilitated by supplies of water obtained from wells deep enough to tap underground sources.

Oases occur in desert areas where water-bearing strata come to the surface. At many points in the Sahara, for instance, water from natural springs or shallow wells permits small-scale irrigation. A high water-table allows the date palm to flourish. This tree provides a valuable food and welcome shade from the burning desert sun. In Fig 87 we can see the barren sands of the Sahara made productive by water from underground sources.

Carboniferous limestone scenery

Carboniferous limestone is a hard grey or white rock formed in seas which existed over 300 million years ago.

It has well-marked joints and bedding planes, the former being arranged roughly at right angles. This rock is composed of calcium carbonate, which is soluble in acid. Rain water is a weak acid, and its slow solution of the limestone gives rise to a most distinctive type of scenery. This scenery is often described as *karst* after an area in Yugoslavia where it is particularly well-developed. In Britain its distinctive features are well displayed in the Mendips and in parts of the Pennines.

Fig 88

Fig 88 is a photograph taken on the flanks of Ingleborough, in the Yorkshire Pennines. It shows, more than 300 metres above sea-level, the plateau surface which is typical of limestone areas in Britain. The soil cover is seldom more than a thin layer and bare rock is frequently exposed. The limestone on the right of the photograph well illustrates the effect of rainwater on this soluble rock. Solution is greatest along the joints which are enlarged into deep, narrow crevices or *grikes*. The blocks left upstanding are called *clints*. The whole feature is an example of a *limestone pavement*, the formation of which is the subject of Fig 89. Note how the clints in the foreground of Fig 88 have been delicately shaped by solution. Such shapes may be recognised in many a park or garden, for weather-worn limestone is much used in rockeries.

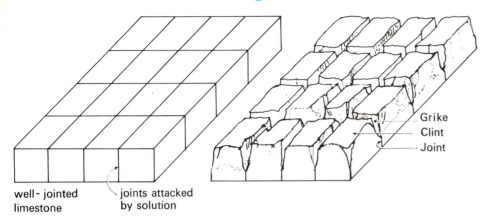

well-jointed limestone

joints attacked by solution

Grike
Clint
Joint

Fig 89

The Ingleborough area provides us with many examples of the characteristic features of karst scenery. The arrangement of the rocks in this area is shown in Fig 90. A thick layer of well-jointed limestone rests horizontally on impermeable rock. It is capped by the great bulk of Ingleborough which is mainly composed of the assorted sedimentary rocks known collectively as the Yoredale Series. Many tiny streams are born on the Yoredales and those to the south-east are collected into Fell Beck which flows onto the limestone.

Fig 90

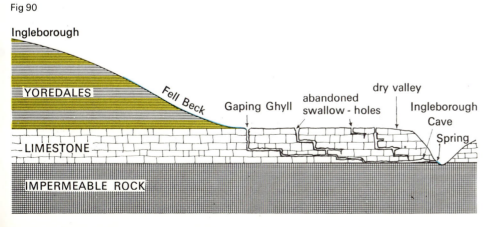

Ingleborough

YOREDALES

Fell Beck

Gaping Ghyll

abandoned swallow-holes

dry valley

Ingleborough Cave

Spring

LIMESTONE

IMPERMEABLE ROCK

Originally Fell Beck continued across the limestone and eroded a valley for itself in the normal way. Slowly, however, the joints over which it flowed were widened as water gradually seeped downwards. With the passage of time, more and more water was able to escape underground by means of these enlarged joints, which are known as *swallow-holes* or *pot-holes*. Eventually the swallow-hole nearest the Yoredales, Gaping Ghyll, was opened out to such an extent that it could accommodate the entire flow of Fell Beck. Its former valley is now dry and marked by a series of abandoned swallow-holes. Fig 91 shows Fell Beck as it disappears from view over the edge of Gaping Ghyll. It tumbles 111 metres in a spectacular underground waterfall before becoming an underground stream which eventually emerges as a spring.

Fig 91

The limestone today is a maze of subterranean passages, such as the one in Fig 92. Underground rivers rush along channels they have carved for themselves by solution and normal stream erosion. Their angular courses reveal the guiding influence of joints and bedding planes. Locally, where erosion has been most effective, these channels open out into extensive caverns. Underground rivers often descend to lower levels as joints are newly opened by solution. Many channels have been abandoned by the rivers which created them, and are completely dry

Fig 92

result, a minute quantity of calcium carbonate is deposited. Slowly, over the years, this deposition builds up into the long slender stalactites which decorate the roof. When drops of water fall on the floor, more calcium carbonate is deposited and the shorter, thicker stalagmites grow upwards. In time stalactites and stalagmites may meet to form a column of calcium carbonate from floor to roof.

Fig 93

except for the steady drip, drip of water from joints in the roof.

Fig 93 is a photograph taken in one of the many caverns in the Ingleborough district. It shows examples of *stalactites* and *stalagmites*. Drops of water seeping through to the roof of the cave are heavily charged with calcium carbonate in solution. As they hang on the roof they lose carbon dioxide to the atmosphere and, as a

Fig 94

Limestone, like chalk, gives rise to a dry landscape. Rivers are few and far between and are only found in the rare valley which is deep enough to reach the low water-table. The limestone valley is typically dry, steep-sided and gorge-like. A small example appears to the left of Fig 88 and a more dramatic example, The Winnats, near Castleton in Derbyshire, is the subject of Fig 94. The steepness of the sides is due to rapid downward erosion when the valley contained a river, and the limited mass wasting in these areas of highly permeable rock.

Fig 95

Occasionally, a valley of similar nature may be produced by the collapse of the roof of an underground watercourse. Trow Gill, Fig 95, is an example from the flanks of Ingleborough.

These limestone valleys may profitably be contrasted with the dry valley in chalk shown in Fig 85.

7
The Work
of Ice

Ice present

A considerable part of the earth's surface is blanketed by thick layers of permanent snow and ice. The greater part of this is found in the polar regions of both hemispheres. Fig 96 is a sketch map of the Arctic and it shows that this area consists mainly of an ocean basin. The intense cold of these high latitudes ensures that the ocean is covered by a crust of ice several metres thick. In winter, this *pack-ice* seals off the Arctic basin and links the U.S.S.R. and North America in its icy grip. In summer it shrinks to the area indicated on the map.

Fig 97

Fig 96

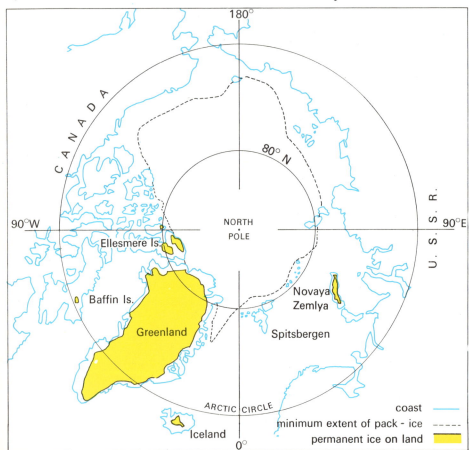

coast ————
minimum extent of pack - ice - - - - -
permanent ice on land ▉▉▉

At this time the pack-ice frequently splits into large floes separated by 'leads' of open water. Drifting with wind and current, these floes are often in collision. Masses of broken ice are driven upwards and freeze into high and tangled ridges. It is under this thick ceiling of pack-ice that nuclear submarines navigate the Arctic waters. On one voyage the U.S. submarine *Skate* surfaced through the ice at latitude 90°N precisely. To reach the same position Robert E. Peary, the first man to gain the Pole, had to sledge for 37 days in severe cold over the broken and irregular surface of the Arctic pack-ice.

Look again at Fig 96 and note that ice is not restricted to the frozen surfaces of the polar sea. Considerable areas of land are also permanently covered by ice, the largest example being that of Greenland. The surface of this vast island is, except for a narrow coastal strip, completely masked by a vast *ice-sheet*. This ice has a different origin. In these cold climates the weak sun of the short Arctic summer has not the power to melt all the annual snowfall. Year by year it accumulates and, by a variety of processes, is gradually converted into solid ice.

The Greenland ice-sheet has been found, by modern echo-sounding techniques, to have a maximum thickness of over 3000 metres. It is thinner towards the margins, where the photograph shown in Fig 97 was taken. Here, the peaks of buried mountains are exposed as *nunataks*. The ice is not stationary, but moves slowly, imperceptibly outwards. In the photograph ice is flowing between the nunataks. It joins the larger mass of ice in the foreground, which is moving from left to right. Eventually it meets the sea and breaks up into icebergs which drift southwards on ocean currents. Before melting in warmer waters, the larger icebergs travel far enough south to become a menace to shipping. The *S.S. Titanic* is, perhaps, the most famous of their many victims.

The Antarctic, unlike the Arctic, is a land mass of continental proportions. It is covered by an ice-sheet which, with an area of 12 million square kilometres, is seven times larger and no less thick than that of Greenland. Echo-sounding has revealed that the ice rests on a rock surface that is broken into what would be islands were the ice to melt. Around the rim of Antarctica, mountains rise above the ice to heights as great as 5139 metres. They include among their number an active volcano in Mount Erebus. This frozen and barren continent is defended in winter by a ring of pack-ice some hundreds of kilometres wide.

Permanent snow and ice in a contrasting environment are shown in Fig 98. This photograph was taken in August and shows part of the Swiss Alps. The loftiest peak is that of the Finsteraarhorn. At these high altitudes temperatures are too low to melt all the snow that falls. Slowly it accumulates and is turned into ice. Snow and ice rest unevenly on this area of jagged relief. Little can cling to the steep rock faces, but in sheltered hollows it collects to great depths. In the top left-hand corner of the photograph we can identify one of these collecting grounds. From here a stream of ice slowly moves downhill under the influence of gravity. This small *glacier* is a tributary to the Aletsch glacier, the longest in Europe, which dominates the left-hand side of the photograph. Slowly, almost imperceptibly, this great mass of ice moves downhill until it wastes away in the higher temperatures of lower levels. The dark lines which streak the broken surface of the glacier are known as *moraines*. They are made up of fragments of rock, weathered from the steep valley sides, which have come to rest on the top of the glacier. They are carried along on the ice and are deposited when it melts.

These same features may also be identified in Fig 106. The glacier to the right of this photograph is the Gorner Glacier, which is born in the shadow of Monte Rosa, one of the highest peaks in the Alps.

The level above which an upland area bears a cover of permanent snow and ice is known as the *snow-line*. It is neatly illustrated in Fig 38. Its altitude varies with latitude. Mount Kenya, an East African peak on the equator, is capped with ice above 5000 metres. In the Alps the snow-line is at about 2700 metres, and in southern Norway the plateau surface is ice-covered above 1400 metres. Britain's highest mountains just fail to rise to the level of permanent ice, although in sheltered hollows on the flanks of Ben Nevis snow occasionally persists throughout the summer.

Fig 98

Ice past

Although Britain today cannot boast any areas of permanent snow and ice, conditions in the recent geological past have been vastly different. Over the last million years or so, for reasons still unknown, average temperatures have shown small but significant variations. With a fall in temperature of just a few degrees Celsius, ice was able to accumulate in Britain's mountain areas just as it does in Alpine regions today. From the mountains of Wales, the Lake District and Scotland, powerful glaciers invaded the lowlands, only to retreat when temperatures fluctuated upwards again. Four times ice advanced and retreated in this way. With the maximum ice advance mountain glaciers merged to form a thick continuous ice-sheet which blanketed the whole of Britain as far south as the line of the river Thames.

The *Ice Age* was not, of course, unique to Britain. Fig 99 shows that Britain's ice cover was only a small part of the vast ice-sheet that covered much of northern Europe and its marginal seas. Similar conditions prevailed in North America. The whole of Canada was buried under an ice-sheet that extended deep into the U.S.A. At this time ice was also more extensive in high mountain ranges such as the Alps and Himalayas.

Those parts of the world which are today burdened with permanent snow and ice are still in the grip of the Ice Age. Britain was finally freed from its clutches a mere 10 000 years ago and its imprint is still fresh upon the landscape. In many areas the tell tale signs of ice erosion and deposition may be readily identified. The scenic beauty of mountain areas, for instance, owes much to the erosive work of ice. The enthusiastic gardener who struggles with a heavy, sticky, boulder-studded soil has the Ice Age to blame for his difficulties.

Glacial erosion

Ice, though slow-moving, has tremendous power to shape and mould the landscape. Upland areas are carved into a set of distinctive landform features by the erosive power of the glaciers they support. Ice on its own does little damage, but it soon equips itself with the tools for the job. It freezes onto the solid rock, especially where well-jointed, and as the glacier moves, blocks are *plucked* away. These rock fragments, and others derived from weathering on the ice-free slopes above, become frozen into the sides and bed of the glacier. As the ice moves slowly downhill, these fragments scratch and scrape away at the sides and floor of the valley, which is greatly enlarged. This process is known as *abrasion*. On exposed rock surfaces we can often still see the scratches or *striations* gouged out by rock fragments held tightly in the grip of a long-vanished glacier.

Fig 100 represents a small feature that is commonly found on the floors of glaciated valleys. It is a *roche moutonnée* and illustrates the two aspects of ice erosion. During the Ice Age an outcrop of resistant rock

Fig 99

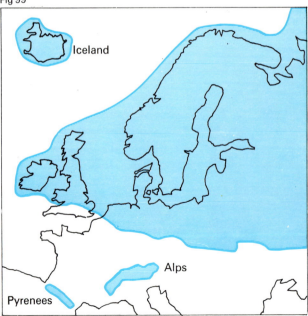

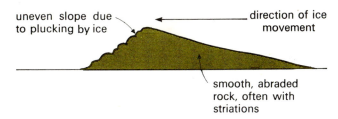

uneven slope due to plucking by ice — direction of ice movement

smooth, abraded rock, often with striations

Fig 100

lay in the path of a glacier moving from right to left. On the upstream side it was smoothed by abrasion but, in its lee, plucking produced the characteristic uneven slope. A larger feature known as a *crag-and-tail* tells us more about ice erosion. The best-known example is in the centre of Edinburgh. Here a crag of tough rock, once the vent of an ancient volcano, lay in the path of the ice. It was big and strong enough to divert the flow of ice and the softer rocks in its lee were largely protected from erosion. Today, the crag is the site of Edinburgh Castle and from the castle, High Street runs down the tail into the city, Fig 101.

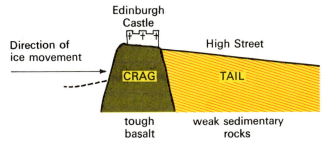

Fig 101

Fig 102 shows one of the distinctive features of a heavily-glaciated upland area. This feature, roughly circular in shape, is usually known as a *cirque*, but it may also be called a *cwm* or *corrie*. On three sides, the rock walls are high and steep. The floor of the cirque has been hollowed out and contains a tarn, from which a small stream flows over the lip of the cirque to the valley below.

Cirques are found perched high above the main valleys of glaciated regions. They have developed from the

Fig 102

sheltered hollows where, during the Ice Age, snow accumulated to great depths. The persistence of this snow encouraged weathering. Summer melt-water and temperatures hovering about freezing-point made freeze-thaw action very effective. The underlying rock was rapidly broken down and washed away. By this means the hollow was enlarged until it was big enough to support its own little glacier. Fig 103 is a cross-section of a cirque at this stage of its development, and an active example may be identified in the top left-hand corner of Fig 98. The plucking action of the glacier keeps the walls steep and causes them to retreat. By this process the cirque is gradually enlarged. The glacier moves out of the cirque with a rotary movement and abrasion hollows out the floor and so creates the rock basin later to be occupied by a tarn.

Fig 103

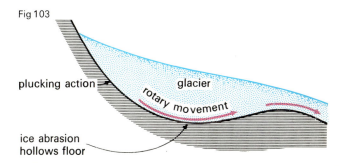

Fig 104

Fig 105

As adjoining cirques grow larger, the land which divides them becomes steadily narrower. The sharp ridge that often separates two cirques is known as an *arête,* and a good example is shown in Fig 104. This is Crib Goch, which provides an exhilarating route to the summit of Snowdon, especially in winter.

In Fig 105 we have an aerial view looking north over the summit of Helvellyn in the Lake District. A series of sharp arêtes reaches out to the east. The most southerly arête is the well-known Striding Edge which forms one wall of the impressive cirque containing Red Tarn.

Fig 106

During the Ice Age the flanks of some mountain masses supported several cirques which bit deeply into the land surface. Snowdon is a case in point. Today, it is mainly composed of a series of arêtes that meet at the summit. In some cases, where glaciation has been more intense, the cirques have eaten further into the mountain and the arêtes have become more severe to form a sharp mountain peak. This feature is known as a *horn* or *pyramidal peak.* Fig 106 gives an unusual view of the best-known example – the Matterhorn on the Swiss–Italian border. We see it in the centre of the photograph pointing steeply upwards at the wing of the aircraft from which the photograph was taken.

Fig 107

In glacial times, ice from several cirques merged into a glacier some 300 metres thick which moved slowly northwards along the pre-existing valley. This glacier eroded deeply into the rock of the valley floor. Fig 108 shows that the pre-glacial valley was converted into a deep, steep-sided trough with a characteristic U-shaped cross-section. High up on the valley sides traces of the pre-glacial valley remain as slopes of more gentle gradient. In Alpine regions, these surfaces, known as *alps*, are more extensive and provide valuable summer grazing for cattle.

A glacier is much less flexible than the river it replaces. It cannot, for instance, easily adapt itself to the interlocking spurs of the normal mountain valley, Fig. 50. Instead it bulldozes its way through and the ends of the spurs are cut away, or *truncated*, leaving the valley straightened. Fig 109 illustrates these changes.

The effect of the tremendous erosive power of ice is best seen in the valleys of glaciated upland areas. Fig 107 gives us an impression of the grandeur of one such valley, in this case the Pass of Llanberis in North Wales. The small size of the stream just visible beyond the road, in the right foreground, is an indication that forces other than running water helped to excavate this huge trough.

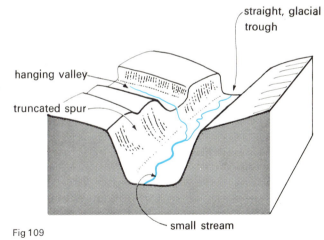

Fig 109

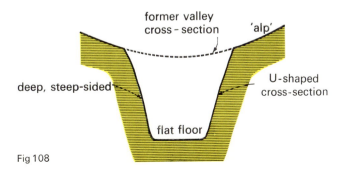

Fig 108

Fig 109 illustrates another distinctive feature of glaciated uplands, the *hanging valley*. During the Ice Age the glacier occupying a main valley was much larger than those in its tributaries. Being larger, it was more powerful and eroded a deeper trough. With the melting of the ice, the tributary valleys were left 'hanging' well above

Fig 110

soil and shaped and polished the rock beneath. Weaker sections were eroded into a multitude of hollows which are today occupied by lakes of all shapes and sizes. In Britain the ice-sheets that spread over the lowlands were relatively small and moved sluggishly. They brought about great modifications of the landscape, but by deposition rather than erosion.

Glacial deposition

The moraines which mark the surface of the glacier in Fig 98 are made up of boulders that are being transported by the slowly-moving ice. These moraines represent only a small fraction of the glacier's total load. Much more material is being carried along unseen at the base of the ice. This is known as *ground moraine* and is composed of rock fragments of all sizes. They range from huge blocks plucked away from the valley floor to the finest rock flour, the product of abrasion. With the melting of the ice, this tremendous load of debris is deposited and leads to great changes in the landscape.

An interesting example of glacial deposition is shown in Fig 111. Here we see a large boulder resting precariously on the plateau surface at Norber near Austwick in west

the floor of the main valley. Fig 110 provides us with an example from Buttermere. The small river tumbles down a steep slope in a series of waterfalls before reaching the level of the lake.

A glacier does not erode uniformly. Where ice is thicker, or rock is weaker, erosion is more effective and long, narrow hollows are quarried from the solid rock of the valley floor. In post-glacial times these hollows fill with water and become lakes. From their shape they are known as *ribbon lakes* and many examples may be found in Britain's glaciated uplands. The depth of these lakes, although often owing something to deposition, is a testimony to the erosive power of ice. Wastwater in the Lake District has a maximum depth of 79 metres. The Loch Ness monster, if such there be, inhabits a glacial ribbon lake no less than 235 metres deep.

The distinctive landscapes of northern Canada and Finland are the result of active erosion by the ice-sheets they bore in glacial times. Ice swept away the mantle of

Fig 111

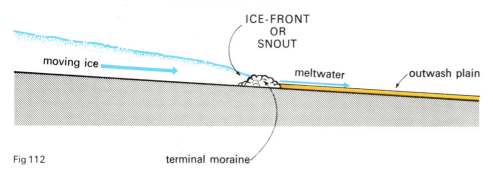

ICE-FRONT
OR
SNOUT

moving ice

meltwater

outwash plain

Fig 112

terminal moraine

many in the district, only had a short ride. It was transported 5 km from its place of origin. In glacial times, of course, ice travelled much greater distances. Erratics of the distinctive granite from Shap in Westmorland can be found in Yorkshire. The rock illustrated in Fig 5 (vi) only occurs naturally in Norway. This particular specimen was picked up by a student on a beach north of Scarborough.

Erratics tell us much about the movement of ice during the Ice Age, but otherwise they are of little significance. Fig 112 will help us to understand the more important forms of glacial deposition. In this diagram we see the ice at the end of its journey. It has reached the point where temperatures are high enough to melt all the ice which arrives. This point is the *snout* of the glacier or the *ice-front* of an ice-sheet. Both types of ice give rise to the same depositional features, but those of the latter are on a much larger scale. Where the ice melts, its load is deposited and accumulates as an unsorted mass of debris known as a *terminal moraine*. Terminal moraines form low, hummocky ridges across the floors of many of the glaciated valleys in upland areas. In Fig 113 we see the example at the head of the Langdale valley. Larger examples are provided by former ice-sheets. In Norfolk the Cromer Ridge, 100 metres high, marks the furthest advance of an ice-sheet that extended down the east coast of England. In northern Germany a terminal moraine, in places over 300 metres high, forms the line of irregular hills known as the Baltic Heights.

Yorkshire. This boulder, which is composed of dark grit, is obviously alien to the limestone area in which it is found. It was brought here by ice and is a striking example of a *glacial erratic*. It must be stressed that the ice did not place the boulder neatly on its three-pronged pedestal. Its elevated position is the result of the lowering of the limestone surface by solution in the time which has elapsed since the Ice Age. This particular erratic, one of

The water from the melting ice-sheet picked up much material as it flowed through the moraine. This material was later deposited to form wide expanses of *outwash plain*. These areas are mainly composed of sands and gravels and today provide poor soils of little agricultural value.

With the gradual rise of temperature which heralded the end of each ice advance, glaciers and ice-sheets began their long, slow retreat. Higher temperatures meant a more rapid thaw, and the ice was not able to travel as far down

Fig 113

direction of
ice movement

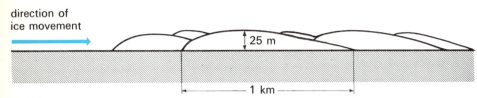

Fig 114

Fig 115

the valley or over the lowlands before wasting away. It must be stressed that the ice did not slip into reverse and drive backwards into the hills. It was the snout or ice-front that moved backwards, thus liberating ever-increasing areas from the burden of ice.

During the period of retreat, large areas of ice became stagnant, and as it melted away its load was laid gently on the land surface. This deposit we know today as *boulder-clay* or *till*. It is composed of stones of various sizes set in a mass of heavy clay. In Fig 123 the brown boulder-clay can readily be distinguished from the white chalk on which it rests. East Anglia and Cheshire are two areas where the pre-glacial landscape has been buried beneath blankets of boulder-clay that are 100 metres thick in places. Soils derived from this deposit are often heavy and hard to plough, but may be of considerable fertility.

The boulder-clay surface is usually level or gently un-dulating, but in some areas it has been moulded into low hills of distinctive shape known as *drumlins*. These features may occur singly or clustered together in swarms, Fig 114. They vary in size but on average are about a kilometre long and 25 metres high. They have a streamlined shape, and the steeper, blunter end faces the direction from which the ice came. The characteristic landscape of drumlin country can be appreciated from Fig 115 which shows a small part of the swarm in the Ribble valley, north of Settle in west Yorkshire.

8
The Work
of the Sea

Fig 116

Fig 117

The sea is a most powerful and effective agent of erosion. The great variety of Britain's coastal scenery is evidence of the work of the sea in all its aspects. The bold cliffs of Land's End, Fig 116, for example, are the result of constant wave attack. Here the land is losing its battle against the sea and is slowly retreating. Elsewhere in quieter corners of coastal waters it is the sea that is in retreat for marine deposition leads to the creation of new land, new coastlines. The conflict between land and sea never ceases and everywhere the coastline is subject to slow but steady change. This chapter will help us to understand the processes of erosion, transportation and deposition which are operating around Britain's coasts, and also help us to explain the varied and often beautiful scenery of our coastline.

Fig 117 reveals the sea in erosive mood. It shows a wave breaking with obvious power on an exposed part of the Cornish coast. The constant repetition of such hammer blows has a destructive effect that is increased when the wave is armed with pebbles and other rock fragments picked up from the shore. Moreover, as the wave breaks against the rock, the air in the cracks and crevices is compressed, and when the water falls back, this compressed air expands with an explosive force sufficient to weaken the toughest rock. The wave in the photograph is

of little more than average size and its erosive power is much less than that of the huge storm-waves sometimes driven against the coast by gale-force winds. There is little that can withstand the fierce and persistent attack of the sea. The most resistant rock, like the strongest defensive wall man can build, will surrender, in time, to its erosive power.

Much of the detail of the coastline is a result of the varying resistance offered by different rocks. Wave attack, like that of all the agents of erosion, is most effective against weaker rocks. Fig 118 is a sketch map of the geology of part of the Dorset coast. Pronounced

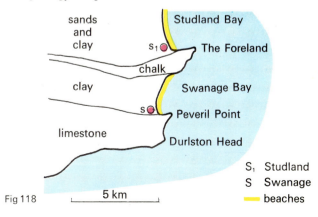

Fig 118

sands and clay

chalk

clay

limestone

Studland Bay

The Foreland

Swanage Bay

Peveril Point

Durlston Head

S_1 Studland
S Swanage
beaches

5 km

Fig 119

Fig 120 (iii) This collapse leads to the formation of a cliff. The fallen material is gradually broken up and removed by the sea, which is then free to continue its assault on the land. The cycle of undercutting, collapse and removal is repeated time and time again and results in the gradual retreat of the cliff, which usually increases in height as it recedes.

Fig 120 (iv) This shows a later stage when the cliffs are well-developed. A considerable area of land has been lost to the sea. Much of this

headlands alternate with wide bays. The softer clays and sands have been scooped out into the smooth bays of Studland and Swanage. Extensive sandy beaches have accumulated in the shallow and sheltered waters at the head of these bays. The headlands, such as The Fore-land, Fig 119, are, however, exposed to the rigours of wave attack and meet the sea in steep cliffs up to 100 metres high.

Cliffs are another result of the erosive work of the sea and their formation is explained with the aid of the diagrams in Fig 120.

Fig 120 (i) A land surface is newly exposed to the sea. Wave attack is restricted to the narrow zone between high and low tide levels.

Fig 120 (ii) In this zone, the breaking waves cut a distinct notch in the land surface. As the notch is extended, the rock above is left unsupported and collapses.

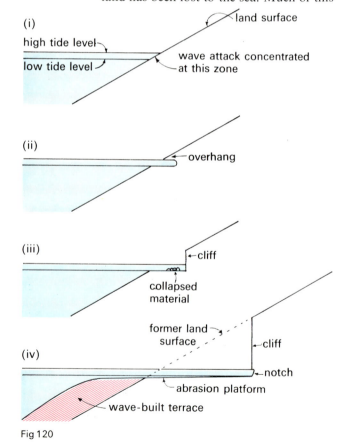

Fig 120

Fig 121

Fig 122

The rate at which cliffs retreat depends largely upon the resistance of the rock under attack. The tough granite of Land's End gives ground slowly. In contrast, the low boulder-clay cliffs of the Holderness area of Yorkshire are retreating at the rapid rate of about 2 metres a year. Since Roman times, this coastline has been pushed back a distance of over 4 km, and many villages have been lost to the sea. Evidence of cliff retreat is provided by Fig 121, taken at Robin Hood's Bay. On the extreme left, part of a road and its bounding wall have collapsed into the sea. A small outbuilding is now perched precariously on the very edge of the cliff. Further retreat will cause its disappearance and menace the row of houses in the top left-hand corner of the photograph.

material has been deposited as a wave-built terrace below low tide level. At this stage cliff retreat will be slow for much of the power of the waves is lost in crossing the extensive wave-cut platform they have created.

Fig. 122 is a view of another part of the Yorkshire coast, in this case the north side of Flamborough Head. Here, the chalk has been 'quarried' into imposing white cliffs which are as impressive as the more famous examples at Dover. Weaker sections of the chalk have been hollowed out by the sea to form two of the many small coves that fret this coastline. In the middle of the photograph more resistant chalk stands out as a small headland that shows evidence of undercutting. The small waves sweep gently inshore across the wide wave-cut platform created by the retreat of the cliffs.

The intricate detail of cliff scenery is an indication of the selective nature of wave attack. Wherever the rock displays the slightest weakness, it is seized upon by waves and eroded with greater vigour. The *caves* in Fig 122 have all been eroded where the chalk is marked by joints or faults which have made the work of the sea just that little bit easier. A line of weakness that runs right through an exposed headland will be excavated from both sides at once. The caves thus formed will eventually meet, leaving the rock above in the form of a *natural arch*. An example of this feature is shown in Fig 123 which was taken from the edge of the cliffs at Flamborough Head. These natural arches eventually collapse leaving an

Fig 123

isolated pinnacle of rock known as a *stack*. Fig 123 may again be consulted for an example.

A more impressive stack is shown in Fig 124, again from the Flamborough Head area. The name 'stack' is given to

Fig 124

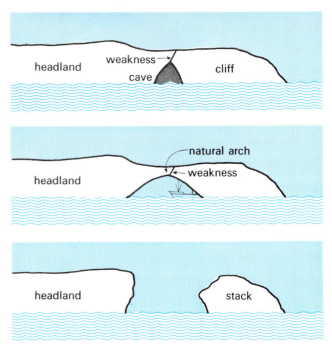

Fig 125

any isolated rock that has been left behind by the retreat of the coastline. They are often a danger to shipping, and large examples often form the foundations of important lighthouses.

Transport and deposition

Waves, as we have seen, are responsible for the erosion of exposed coastlines and the sea uses this same tool for its work of transportation and deposition. Waves breaking on the beach cause a movement of sand and shingle which leads to the formation of new features of coastal scenery.

Study Fig 126. It shows a regular procession of waves rolling towards the shore. One by one, in rapid succession, these waves will break on the wide beach. Note that they are not coming in at right angles to the beach, but approaching it at an oblique angle. The force that forms

Fig 126

and drives the waves is the wind. As the wind is everywhere variable in direction, it follows that waves may approach the beach from any angle. For any particular beach, however, there is usually one direction from which the waves approach most frequently or most forcibly.

In Fig 127 it is assumed that the waves approached most frequently from the south-west. As each wave breaks it carries material obliquely up the beach. As the backwash returns to the sea, it will do so under the influence of gravity and will flow down the beach at right angles. Thus a pebble or a grain of sand that was picked up at A will come to rest at B, there to await the next co-operative

wave which will, in the same way, take it further along the beach to C. By this process, which is known as *longshore drift*, sand and pebbles are slowly but steadily moving along the shore. The occasional waves that approach from the south-east will temporarily reverse the direction of movement, but the greater frequency of waves from the south-west will ensure that longshore drift is in the direction indicated on the diagram.

Longshore drift affects many of the beaches around the coasts of Britain. An example is that at Eastbourne, which is shown in Fig 128. Here, at this south coast resort, longshore drift is from south-west to north-east, that is

Fig 128

from bottom to top of the photograph. Substantial *groynes* have been constructed at right angles to the promenade in order to check the movement of beach material. The piling up of sand on the near side of each groyne is evidence of the drift of material along the beach.

Longshore drift leads to the formation of several distinctive features of coastal scenery. In Fig 129 material on the beach at A is moving from left to right. Longshore drift

Fig 127

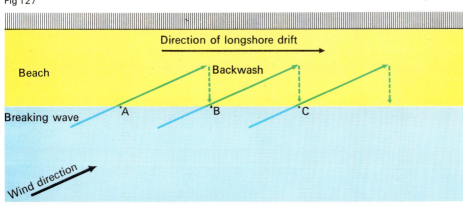

Direction of longshore drift

Beach

Backwash

Breaking wave

A   B   C

Wind direction

Longshore drift

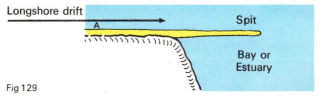

A   Spit

Bay or Estuary

Fig 129

Fig 130

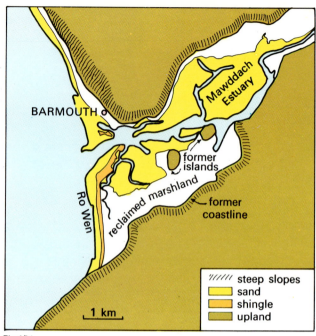

Fig 131

does not cling to the shore when the direction of the coastline changes, but causes the beach to extend into the open water of the bay or estuary, as a *spit*. Fig 130 shows Spurn Head which is a fine example. This thin ribbon of sand and shingle has grown out into the Humber Estuary under the influence of longshore drift which, on this part of the Yorkshire coast, is from north to south. Like many spits the tip of Spurn Head has been bent into a hook shape by the influence of the waves.

Fig 131 provides us with another example of a spit. On this part of the Welsh coast longshore drift is from south to north and the shingle spit known as **Ro Wen** has grown northwards into the estuary of the Afon Mawddach. Indeed, it is only tidal scour and the outflow of river water that has prevented the estuary from being sealed off completely. The growth of Ro Wen, by creating sheltered waters to the south of the estuary, has encouraged the development of new land by river deposition.

Spits are a common feature of Britain's coastal scenery. In your Atlas turn to the map of East Anglia, and note how the River Alde has been diverted by the southward growth of the large spit of Orford Ness. A little way to the north, a spit is the site of the town of Great Yarmouth, and here again a river, in this case the Yare, has been diverted to the south. Look again at Fig 126, and identify on the photograph the effects of longshore drift at Seaton in Devon.

Fig 132

Occasionally, longshore drift may cause a beach to grow right across the mouth of a quiet bay, creating a lake or lagoon (*haff*) in the process. Such a beach is known as a *bar* (or *nehrung*), and an example from Slapton Sands, Devon, is shown in Fig 132.

Fig 133

A third feature formed as a result of longshore drift is illustrated in Fig 133. The 'island' of Portland is no longer surrounded by water for it has been linked to the mainland by the growth of Chesil Beach. In a case such as this, the beach is known as a *tombolo*. A neat example, set in the waters of a Scottish Loch, is shown in Fig 134.

Fig 134

Types of coastline

It only takes a glance at a map of the world to convince us of the bewildering complexity of the world's coastlines. On close examination, however, we can recognise a number of patterns which are repeated in widely separated parts of the globe. One of these patterns is illustrated by Fig 135 which shows part of the coast of Devon and Cornwall. The main feature of this type of coastline is the numerous estuaries of distinctive shape known as *rias*. Coastlines of this type may also be identified in Brittany, north-west Spain and south-west Ireland.

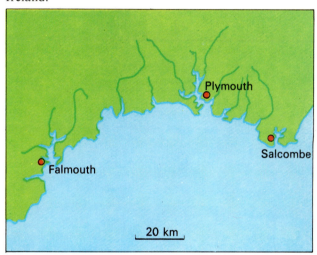

Fig 135

A typical ria is shown in Fig 136. This is the Kingsbridge Estuary and the settlement that clings to the steep hillside on the left is part of the small Devonshire town of Salcombe. Follow the course of the estuary as, in this upland area, it extends far inland. This is no easy task for its course is of a winding nature, and the eye is easily diverted into one or more of the estuary's several branches. The depth of water gradually decreases as the ria narrows inland, eventually to merge into the river at its head.

Fig 136

A coastal feature on a scale such as this is too large to be explained as the result of marine erosion. To understand its formation we need to remember that, as explained in Chapter 3, sea-level is not constant, but subject to considerable variation in relation to the land surface.

Study Fig 137. In (i) a stream with several tributaries winds its way down to the sea. Imagine a rise of say 50 metres in sea-level. The river valley will be 'drowned'. The old 50-metre contour will become the new coastline as a long, many-branched arm of the sea – a new ria – is created, Fig 137 (ii). This submergence will expose a new land surface to wave attack, and we can see evidence of this in the typical cliff scenery in the foreground of Fig 136.

Fig 138

Fig 138 shows part of the coast of northern Yugoslavia and well illustrates the main features of the *Dalmatian* type of coastline. Here the mainland is defended by lines of long, narrow islands. This type of coastline, like the ria, is the result of a substantial rise in sea-level. Fig 139 (i) shows an upland area where the grain of the land is parallel to the sea. Pronounced ridges alternate with low-lying vales. When sea level rises (ii), the low land is drowned and the ridges become islands.

Fig 137

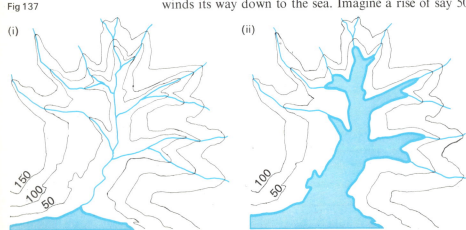

Fig 139

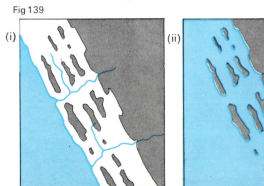

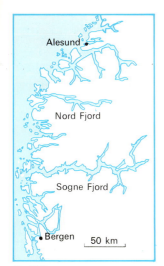

Fig 140

A rise in sea-level has played a supporting role in the formation of a third distinctive pattern – that of the *fjord* coastline. This type is found in British Columbia, Southern Chile, South Island New Zealand and among the sea lochs of Western Scotland. It is best developed, however, in Norway and Fig 140 shows a small section of that country's extensive coastline. From this map we can see that many straight and narrow fjords bite deeply into the land. The head of Sogne Fjord, for example, is 183 kilometres from the open sea. The water in these fjords is often of great depth. Sogne Fjord may again be quoted as an example. Here a depth of 1219 metres is recorded many kilometres from the open sea. Surprisingly, perhaps, the shallowest part of the fjord, the *threshold*, is near its mouth (Fig 141).

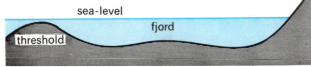

Fig 141

Fig 142 is a view inland along a typical Norwegian fjord. The straightness of its sides; their steepness and towering heights suggest a glaciated valley and fjords are, indeed, due mainly to the work of ice. During the Ice Age, glaciers

Fig 142

descended from the high plateau of Norway and followed pre-existing river valleys down to the sea. These glaciers were so powerful that they gouged out their valleys to depths well below sea-level. When the glacier reached the sea it would float in the dense salt water and eventually break up and drift away as icebergs. Thus at this point, which corresponds to the threshold, erosion would cease.

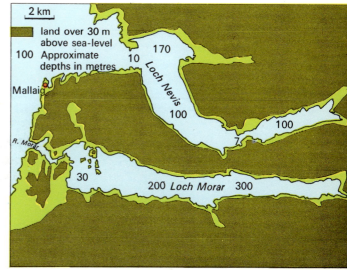

Fig 143

In areas where glaciation was less intense, some glaciers did not quite reach the sea. The troughs they excavated were separated from the sea by a low barrier of rock. Here a rise in sea-level, which in normal circumstances would merely increase the depth of the fjord, is responsible for their formation. Loch Nevis, which is shown in Fig 143, is an example. Note the shallowness of the threshold and the great depths recorded a little way inland. On the same map we can see that a rise in sea-level of 30 metres would convert the fresh-water Loch Morar into a salt-water fjord with a well-marked threshold, the higher parts of which would stand out as islands.

The effects of submergence are not restricted to upland areas. A rise in sea-level on a lowland coast will bring about marked changes and lead to the formation of a distinctive coastline, known, from its origin, as the *submerged lowland* type. Fig 144 gives an example. Sluggish streams flowing across the lowland of Essex and Suffolk have been converted into wide estuaries that extend well inland. Tidal creeks are a common feature. They often encircle patches of land to form low islands. The shallow estuaries are rapidly being filled by river deposition, and marshland, often reclaimed for agriculture, covers extensive areas. The coast of north-west England with the wide estuaries of the Ribble, Mersey and Dee may be quoted as another example.

Fig 144

Submergence has been the dominant factor in the formation of the greater part of the world's coastlines. This is because, in geologically recent times, sea-level has been steadily rising as a result of the long, slow thaw of the world's ice-sheets. It is, in fact, still doing so at a rate of 100 mm every century. Sea-level, however, may fall as well as rise. This change need not imply a reduction of the amount of water in the oceans, for it may be brought about when the land is uplifted by geological processes.

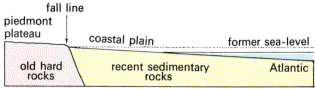

Fig 145

A fall in sea-level relative to the land causes part of the sea floor to emerge and become dry land. When the sea is bounded by an area of pronounced relief, a fall in sea-level has little effect. Indeed, these *emerged upland* coasts are often only detected by the presence of raised beaches (page 19). A fall of sea-level on a lowland coast is of much greater significance for extensive areas of the sea floor are added to the world's land area. The coastlands of the south-eastern region of the U.S.A. provide us with an example of this *emerged lowland* type of coastline. Fig 145 is a simple cross-section of the area. The recent sedimentary rocks and the gentle seaward slope of the coastal plain is evidence of its submarine origin. The coastline is rapidly being modified by marine and river deposition in its shallow waters. The position of the old coastline is marked by a distinct break of slope known as the 'fall line'. This feature is aptly named for at this point rivers tumble from old to new land in a series of waterfalls.

A distinctive type of coastline occurs off Australia's Queensland coast. Here, from beyond Cape York almost as far south as the Tropic of Capricorn, the coastal waters are marked by a series of reefs known collectively as the Great Barrier Reef. This feature is over 2000 km long and its distance from the mainland varies from 30 km to 240 km. These reefs are alive with primitive marine organisms called coral polyps. These tiny creatures extract calcium carbonate from sea water and use it to build the cup-like cavities in which they live. When one polyp dies, its place is taken by another which adds its contribution of coral limestone to the growing reef. Most coral polyps live in colonies and Fig 146, which was taken on the Great Barrier Reef, indicates the great variety of forms and colour that these colonies take.

Fig 146

it is separated by the shallowest and narrowest of lagoons. The most interesting result of coral growth, however, is the *atoll*. An example is Aldabra Island in the Indian Ocean, Fig 147. The atoll consists of a ring or horse-shoe of low coral islands, sometimes unbroken, which encloses a shallow lagoon. These islands are often covered with coral sand which allows the sea-borne seeds of the coconut palm to take root and flourish.

Coral is of limited distribution. As the polyps can only flourish in water with temperatures higher than 20°C, they are restricted to tropical seas. Examples are particularly numerous in the western Pacific, but they may also be found in parts of the Indian Ocean and in the warm waters off the islands in the Caribbean Sea. Coral growth also demands sunlit salt water which is free from silt. A significant point is that the polyps cannot survive at depths greater than about 30 metres. The upward growth of coral is limited by the level of the sea.

The origin of these coral reefs has been a source of speculation and controversy for many years. The formation of fringing reefs presents no problems, for they simply grow in the shallow waters of tropical coasts.

Coral gives rise to other coastal features. It is often found as a *fringing reef* hugging the coastline from which

Fig 147

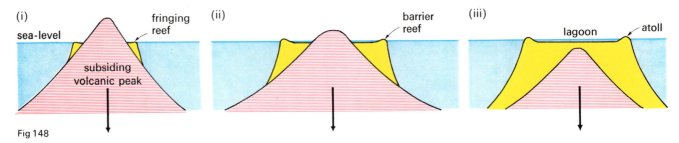

Fig 148

But barrier reefs and atolls rise steeply from depths at which coral cannot grow. Charles Darwin, as long ago as 1842, was the first to suggest a possible explanation. Darwin linked all three types of coral in a theory that was based on an assumption of considerable subsidence of land areas. His theory is explained with the aid of Fig 148. In (i) a volcanic peak rises above sea-level. A fringing reef develops in the shallow coastal waters. The peak is slowly sinking, but the coral reef grows upwards and outwards keeping pace with the subsidence, to become a barrier reef as in (ii). Fragments of coral are broken off by breaking waves and hurled over the reef to form the level floor of the lagoon. With continued subsidence the peak disappears below sea-level (iii) and the reef remains in the circular shape of the typical atoll.

Other theories have since been put forward, but the great depth of coral rock revealed in recent borings lends great support to Darwin's ideas. Drilling on Eniwetok, a Pacific atoll, showed it to be composed of over 1200 metres of coral resting on volcanic rock. The formation of the Great Barrier Reef is also thought to be due to extensive subsidence, for this part of the coast of Queensland has been greatly downfaulted.

As a footnote, it may be added that the world's coral coasts are in danger. Parts of the western Pacific are affected by an increase in the population of the crown of thorns starfish which is reaching plague proportions. These voracious creatures feed on the coral polyps, and without the constant additions made by living coral, the reefs are slowly being destroyed by wave action.

9
The Work of the Wind

We have only to observe the wind eddying around a corner of the playground to appreciate its ability to transport loose material lying on the surface. Fine particles of dust are lifted high into the air, and grains of sand and grit are sent scurrying over the asphalt. The more violent the wind the greater the movement. With a strong wind blowing, even grains of sand may be lifted above the surface.

The farmers of the English Fenlands well appreciate the power of the wind. In a dry spring, before the growing crops are high enough to bind the soil, a strong wind may lift and remove the finest particles of this fertile land. The farmer stands helpless as he watches the best of his soil being swept from his fields. This lowering of the land surface is known as *deflation*, and in cultivated areas it can lead to serious erosion of the soil. In the early 1930's unwise cultivation of the dry lands of the American states of Kansas and Oklahoma led to the creation of a wasteland known as the Dust Bowl. Farmland over a wide area was destroyed as strong winds removed the soil.

Even the finest particles lifted by deflation and transported away by the wind must eventually be deposited. This wind-borne dust may be added to the soil mantle or lost to the sea. In some areas, where conditions are favourable, it has accumulated to great depths. Parts of north-west China, especially within the great bend of the Hwang Ho, are covered with a fine-textured deposit known as *loess*. Brought by the wind from the arid wastes of the Gobi desert, this deposit, which is over 100 metres thick in places, provides a soil of high fertility. In some districts it also provides cheap living accommodation. Cave-like dwellings are excavated in the loess which is tightly bound by the roots of growing vegetation. Chimneys commonly protrude into the fields above.

Loess also occurs over much of the mid-west of the U.S.A. and in Europe in a belt from the Paris Basin deep into the U.S.S.R. In these areas the loess is not derived

Fig 149

from the dusts of the deserts, but from the finest products of ice erosion. In glacial times, winds blowing over the unsorted debris left by the ice-sheets selected the finest material and transported it well to the south. As in China, the loess of the U.S.A. and Europe provides a most productive soil.

Other aspects of the work of the wind may be observed in many of our coastal districts. Look at Fig 149. Here, on the Lancashire coast near Southport, a wide expanse of gently shelving beach is exposed at low tide. On a hot day the surface sand dries out and a strong breeze from the sea sends it streaming across the beach. Any obstacle in the path of the wind reduces its speed and sand is deposited. Such an obstacle is shown in the middle of the photograph. A patch of old turf, little higher than a cigarette lighter, interrupts the smooth flow of the wind and causes sand to accumulate as a tiny dune. In this way are sand dunes born. Even the largest examples may have such small beginnings for they grow by the constant addition of wind-borne sand.

The new-born dune seen above was destined not to grow for it was destroyed by the incoming tide. On this same extensive beach, safely beyond the reach of the highest tide, a continuous line of dunes has been built up. In Fig 150 we have a closer view of this jumbled and irregular

Fig 150

mass of sand which rises to heights of over 15 metres. Note the partial cover of tough marram grass. This protects the dunes from the full power of the wind. Without the support of growing vegetation, dunes slowly migrate as sand is blown over the crest by the prevailing wind. This is illustrated in Fig 151.

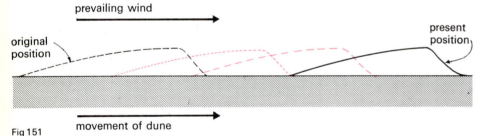

Fig 151

Desert landscapes

It is to the arid regions of the world that we must turn for the most dramatic examples of the work of the wind. In desert areas of high temperatures and low rainfall, weathering produces a large supply of fine rock fragments. In places, sand accumulates to great depths and completely obscures the underlying rock surface. One such area is shown in Fig 152. Here in the great sand sea or *erg* of Arabia the sand has been whipped up into a sea of huge dunes that are often over 100 metres high.

Fig 152

Unprotected by vegetation, these dunes move freely at the dictate of the wind.

The erg is not always the jumbled and chaotic mass of sand shown in the picture. The sand, for instance, may lie in level and monotonous sheets. Deflation may clear the sand from small areas to form shallow depressions known as *deflation hollows*. These, if they extend down to the water-table, sometimes give rise to oases. Often the sand is moulded into dunes of distinctive shape. *Seif* dunes, for example, are long narrow ridges of sand. They occur in groups parallel to the direction of the prevailing wind. Another distinctive type of dune – the *barchan* – is shown in Fig 153. It has a steep, crescent-shaped leeward face, which is extended well forward at the sides.

Fig 153

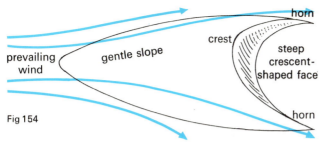

prevailing wind — gentle slope — crest — horn — horn — steep crescent-shaped face

Fig 154

Fig 154 emphasises these characteristics. As the highest part of the dune is at the centre of the crest, the sand-laden wind is diverted to the sides, and hence the barchan's pronounced 'horns'. Barchans occur singly or in swarms, and with a strong and constant prevailing wind behind them, they may achieve speeds of 15 metres a year as they 'march' across the desert floor.

Although the erg is, perhaps, the best known type of desert landscape, it is by no means the most extensive. In fact, sand covers only about 20% of the area of the hot desert regions of the world. Over vast expanses of the Sahara, for instance, the desert surface is a loose and uneven pavement of boulders and pebbles known as the *reg*. Other areas—the *hamada*—are composed of level layers of polished rock. Fig 155 was taken over the south Negev area of Israel and shows a mountain type of desert. It is a stark, barren waste of varied relief.

Fig 155

Exposed rock surfaces in desert regions are targets for the erosive work of the wind. Look at Fig 156 which shows rock formations near the oasis of Al Hassa in Saudi Arabia. On a still day the desert sands rest quietly in the foreground, but with a wind blowing they are whipped up and hurled against the pinnacles of rock. This 'sand blast' effect is known as *wind abrasion*, and its destructive effect can be appreciated from the photograph.

Fig 156

Sand is seldom lifted more than a metre above the surface and abrasion is most effective near ground level. This leads to undercutting, and the formation of *mushroom rocks*. A fine example can be seen on the skyline at the left of the photograph. Soon, this 'mushroom' will topple as abrasion cuts through the 'stalk'. The wind is highly selective in its erosive work, and picks out any weakness with great delicacy. On the rock face in the centre of the photograph, we can see evidence of this. The less resistant bands of rock have been eroded to a greater extent than the tougher strata. These effects are illustrated diagrammatically in Fig 157.

Fig 157

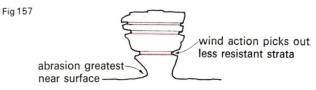

wind action picks out less resistant strata

abrasion greatest near surface

Chapter 9

Water in the desert

Explorers journeying deep into the Sahara have discovered many caves that are adorned with paintings and engravings made by long-vanished artists. These paintings, well preserved in the dry desert air, depict giraffe, antelope and other animals that were the quarry of pre-historic hunting peoples. These animals are herbivorous and today are found only on the grasslands hundreds of kilometres to the south. If, long ago, they flourished in what is now the arid Sahara, the climate of those times must have been vastly different. In particular, rainfall must have been much greater to support the vegetation on which the animals depended for food.

A former climate of heavier rainfall would explain the formation of the *wadi* shown in Fig 158. In many desert areas the land surface is gashed by these deep and gorge-like valleys. The wind has not the power for erosion on this scale. Wadis could only have been formed when rainfall was sufficient to support active erosion by running water.

Even today water is of great significance in modelling the desert landscape. Rain may be rare, but it usually occurs in short downpours of great intensity. A year's average rain may fall in the space of an hour or two. When this happens, dry wadis become raging torrents with a suddenness that has brought death by drowning to unsuspecting desert travellers. These *stream floods* seldom last long or travel far, for water is rapidly lost by evaporation and percolation. They have a short life but

Fig 158

an active one. Armed with a heavy load of rock debris, they erode the wadi floor. The load is transported to lower levels and spread over the desert surface. This flushing of the sand-choked wadis exposes fresh rock surfaces to the destructive effects of weathering and wind action.

In better watered parts of the desert, rainfall may be sufficient to permit streams to flow. These streams are usually highly seasonal and flow only short distances. They soon succumb to evaporation and percolation, and often terminate in brackish lakes or salt flats. The heavier rainfall of the desert margins often supports streams which flow into the desert, but they usually meet a similar fate. Rivers like the Nile and the Tigris-Euphrates which succeed in crossing the desert and reaching the sea are rare indeed.

10 Lakes

A lake may be defined simply as any hollow or basin in the surface of the earth where water has accumulated. Lakes are common features of many landscapes, and occur in infinite variety. In size they range from the merest pond to the giant Caspian Sea for which, with its surface area of 437 000 square kilometres, the name sea is fully justified. The shallowest lakes are only a centimetre or so deep, but at the other end of the scale, Siberia's Lake Baikal has a depth of 1940 metres.

Under normal circumstances lake basins are filled by rain and inflowing rivers. At the lowest point of the rim the water overflows as a river which journeys on to the sea. In arid climates, however, there may not be an outflowing river. In these areas of *inland drainage* the addition made by rivers is balanced by the loss due to a high rate of evaporation. Such lakes, and Great Salt Lake in Utah, is a good example, have a high salinity which prevents their waters being used for drinking or irrigation. Lakes in areas of highly seasonal rainfall

Fig 159

often vary greatly in size. Lake Chad, for example, often has a surface area of 22 000 square kilometres or more, but in the dry season it shrinks to the merest shadow of its wet season glory.

A lake is only a temporary feature of the landscape. Natural processes result in its disappearance in what, in geological terms, is a very short period of time. Fig 159 will help us to appreciate these processes. It is a view of the English Lake District looking over Buttermere to Crummock Water, with Loweswater just visible in the distance. The stream that flows into the head of Buttermere is, as normal, carrying a load of sediment. When it meets the still waters of the lake, its speed is checked and its load deposited. This deposition has led to the creation of a considerable area of new land. We can glimpse a corner of it over the shoulder of the rocky crag in the left foreground of the photograph. Buttermere is being filled from other directions. On its eastern shore can be seen a small delta that is growing out into the lake. The level land between Buttermere and Crummock Water is the surface of a larger delta, the growth of which has divided a pre-existing body of water into the two lakes we see today. Erosion as well as deposition helps in the elimination of lakes. The outlet of Crummock Water is gradually being worn down by river erosion, and this lowers the level of water in the lake and so reduces its surface area.

Borrowdale, shown in Fig 160, is an example where this process of elimination has been completed. The level floor of this Lake District valley is the site of a former lake, created during the Ice Age, and removed from the landscape in the short time that has since elapsed.

Of the many ways in which lakes may be formed, several have been described in earlier chapters. It was, for example, the erosive power of a valley glacier which excavated the basin now occupied by Buttermere and Crummock Water. The tarns that occupy the hollowed-out floors of mountain cirques are also the result of

Fig 160

Fig 161

glacial erosion. Mention has also been made of the innumerable lakes caused by the slow passage of vast ice-sheets over the shield lands of Canada and Scandinavia.

Deposition is another process that frequently leads to the formation of lakes. The familiar ox-bow, for example, is caused by river deposition as it seals off an old meander. A moraine put down by a retreating glacier may pond back a river or, as is often the case, increase the depth of water in an eroded basin. The hollows in unevenly deposited boulder clay are frequently occupied by small lakes such as the *meres* of north-west Cheshire. In Chapter 8 we saw how longshore drift has created the lake at Slapton in Devon.

Lakes may be due to the damming of a valley by a lava flow. A more remarkable type of lake due to vulcanicity is shown in Fig 161. This is Crater Lake, Oregon, and, as its name suggests, it occupies the huge crater of an extinct volcano. It is nearly 10 kilometres across and the island is a subsidiary volcanic cone.

The uneven floor of a rift valley is an obvious site for a lake. Look again at Fig 32 (page 24) and note the strings of major lakes that occur in both arms of the great African Rift. Incidentally, Lake Victoria, the lake shown on that map, is not due to faulting but to the gentle sagging of the earth's crust.

Lakes are useful to man in many ways. They are natural storage basins for water which is a vital resource for domestic and industrial purposes, for irrigation and for the generation of hydro-electric power. Lakes are useful in other ways. They have an amenity value that greatly increases the tourist appeal of such areas as the Lake

Fig 162

District and the Central Plateau of Switzerland. Some lakes have a fish population large enough to support fishing industries of considerable local importance. The Caspian Sea, for example, is the home of the sturgeon, from the roe of which caviar is produced. Large lakes provide opportunities for cheap water transport. The value of such facilities provided by the Great Lakes of North America would be hard to exaggerate.

Man finds lakes such a valuable asset that he frequently copies nature by depositing a dam across a valley thus creating an artificial lake or reservoir. A small example is shown in Fig 162. This is Clatworthy Reservoir on the river Tone in Somerset which is a source of water for the town of Taunton.

The lake in Fig 105 is a case where man has co-operated with nature. The depth of a glacial ribbon lake has been increased by a dam, and the resultant Lake Thirlmere supplies water to Manchester.

The U.S.A. provides many examples of large reservoirs constructed with several purposes in mind. Lake Roosevelt is a man-made lake held up by the Grand Coulee Dam. Besides helping to regulate the flow of the Columbia river, it provides water for irrigation and is used to generate over 2 million kilowatts of electricity annually.

11 Planet Earth

A glance at Fig 2 is a reminder that the earth, our home, is a planet, one of nine which orbit around their master star – the sun. Fig 163 shows diagrammatically that planet earth is subject to two distinct motions. It is constantly spinning like a top on an axis that extends from North to South Poles. The time taken for each rotation we know as a day and this, for convenience, is divided into 24 equal parts known as hours. The constantly spinning earth follows a slightly elliptical orbit around the sun. The distance between our planet and its star varies between 147 million kilometres and 152 million kilometres. The time the earth takes for this journey around the sun is known as a year. In this time the earth rotates on its axis just over $365\frac{1}{4}$ times.

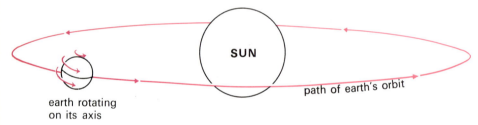

earth rotating
on its axis

Fig 163

The earth's planetary movements give rise to important effects that greatly influence our daily lives. The effects will be considered below.

Day and Night

Fig 164 views the earth from above the North Pole. It shows how the parallel rays of light from the distant sun can bring daylight to only half the earth's surface at any one time. The other half, shaded from the sun, experiences night. As the earth is rotating on its axis, the area illuminated by the sun is constantly changing. With the aid of Fig 164 let us follow the progress of a representative point P on the earth's surface. At P_1 it is in darkness, but when the earth's rotation has

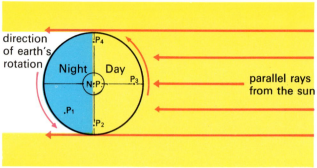

direction of earth's rotation

Night | Day

parallel rays from the sun

Fig 164

brought it round to P_2, the first light of a new day is visible above the eastern horizon. Gradually, as the earth rotates, the sun appears to climb higher in the sky reaching its maximum elevation at P_3. Then it apparently sinks slowly in the west until at P_4 the point passes once more into night.

Time

In Fig 164, P_3 is a position of special significance. It represents noon, the moment when the sun appears to reach its highest point in the sky. Noon is a convenient basis for the measurement of time. It occurs at the same moment at all points on a particular meridian of longitude. The rotation of the earth brings successive meridians to the noon position and thus each line of longitude has its own noon, and hence its own *local time*. Every 24 hours, the earth rotates through 360°. In one hour, therefore, it rotates through 15° of longitude, and in 4 minutes through 1°. Two places that differ in longitude by 30° will have a difference in local time of two hours. Radio and television provide us with good illustrations of time differences over the world. It is early morning when we hear a cricket commentator describe the last over of a day's play in a Test Match in Australia. Pictures of the splashdown of a spacecraft in the first light of a Pacific dawn may appear on our TV screens, via a satellite, in the afternoon of the same day.

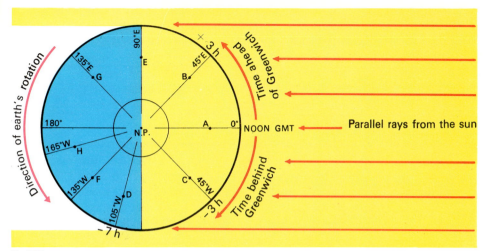

Fig 165

The Greenwich meridian, longitude 0°, is taken as the standard for the comparison of local times. In Fig 165, point A is on this meridian, and the time is noon. Point B is on longitude 45°E. It passed the noon position three hours ago, and its local time is that amount ahead of Greenwich Mean Time, i.e. 15.00. Point C, on the other hand, at longitude 45°W, has not yet reached the noon position. It will take another 3 hours to get there, and the local time is 09.00. Point D, on latitude 105°W, provides us with a further example. 105° is equivalent to a time difference of 7 hours and, being west, it is behind G.M.T. and local time is, therefore, 05.00. Points E, F, G, H, have been included in Fig 165 so that you may test your skill by calculating their local times when it is noon at Greenwich.

If we remember that 1° of longitude represents a time difference of 4 minutes we can easily calculate the local time on each individual meridian. On longitude 17°E, for example, local time will be 1 hour 8 minutes ahead of G.M.T. Similarly, local time at 78°W will be 5 hours 12 minutes behind G.M.T. What will it be at longitude 123°W?

A comparison of local time with that of Greenwich enables longitude to be calculated with ease. Suppose for instance, that a ship's navigating officer notes that local time is 5 hours behind G.M.T. A time difference of 5 hours is equivalent to 75° of longitude, and as his local time is behind that of Greenwich, his longitude must be west. Thus his longitude is 75°W. Using this method, calculate the longitude of an observer when local time is (a) 6 hours 30 minutes behind, and (b) 3 hours 15 minutes ahead of G.M.T.

Let us return to Fig 165 for a moment and consider the odd case of longitude 180° which is both east and west of Greenwich. If we calculate its local time we find that it is either 12 hours behind or 12 hours ahead of Greenwich. Thus there is a time difference of 24 hours, or one whole day, between two adjacent points that lie on opposite sides of longitude 180°. It is necessary to adjust the calendar to compensate for this time difference. On a journey westwards across this meridian, the traveller loses a day: Tuesday, for example, is followed by Thursday. Travelling in the opposite direction, a day is gained: Wednesday, for instance, is followed by another Wednesday before Thursday dawns. The line where these changes are made is the *International Date Line* shown in Fig 166. For most of its length it closely follows the 180° meridian, but deviates slightly to avoid inhabited areas such as Eastern Siberia and the Aleutian and Fiji island groups.

We have now seen above that local time varies by 4 minutes for every degree of longitude. Noon by the sun is, for instance, 24 minutes later in Cornwall than it is in Kent. If every place kept to its own local time, great confusion would be caused, and travellers would have to adjust their watches every few kilometres or so. To overcome this problem, the world is divided into time-zones, each about the width of 15° of longitude, and the time on a central meridian is taken as *standard time*. The U.S.A. is divided into the four zones shown on Fig 167. Each zone takes as standard the time appropriate to the

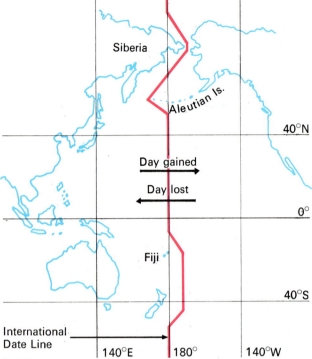

Fig 166

meridian indicated. These times are behind G.M.T. by the number of hours shown on the map. Incidentally, it must always be borne in mind that the time shown on

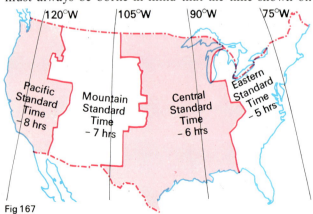

Fig 167

your watch, even if accurate, is not necessarily G.M.T. In order to keep in step with most of Europe, Britain adopted *British Standard Time*, which, one hour ahead of Greenwich, is the standard used by most of our continental neighbours. This decision has since been changed and in 1971 Britain reverted to G.M.T. for the winter months, at least.

Variations in the length of day and night

You will doubtless have noticed that the axis on which the classroom globe rotates is not vertical. Instead the globe is supported in such a way that it spins on a tilted axis. The angle of tilt is $23\frac{1}{2}°$, and this angle is maintained by the earth at all stages of its journey around the sun. This is illustrated by Fig 168 where the earth is shown at four significant points in its orbit – the spring and autumn equinoxes and the solstices of summer and winter. The inclination of the earth's axis has certain consequences of great importance. One of these is the variation in the length of day and night over the year. In high latitudes this produces the marked contrast between the long hours of summer daylight and the short days of winter. In order that we may understand this variation let us follow the earth on one of its orbits. Only the situation in the northern hemisphere is considered. Directly opposite conditions prevail south of the equator.

The winter solstice is a good starting point. It is illustrated diagrammatically in Fig 169. At this position, the inclination of the earth's axis is such that the sun at noon is vertically overhead not on the equator but on the Tropic of Capricorn, $(23\frac{1}{2}°S)$. The northern hemisphere is tilted away from the sun. Night follows day as the earth rotates on its axis. It will be appreciated from Fig 169 that on the equator day and night are of equal length, but everywhere north of this line, there is less than 12 hours of daylight. The higher the latitude, the shorter the day. The earth's rotation spins London through more than sixteen hours of darkness and less than eight hours of daylight. Places north of the Arctic Circle do not

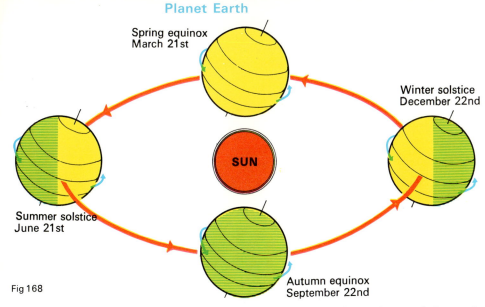

Fig 168

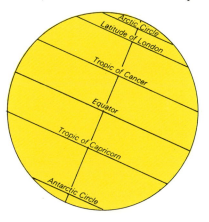

Fig 170

overhead at the equator and as the earth rotates both hemispheres receive equal illumination. At the equinox, day and night are both of 12 hours duration at every point on the earth's surface.

As the earth continues on its journey, the northern hemisphere becomes increasingly tilted towards the sun. The vertical noonday sun achieves steadily higher latitudes in the northern hemisphere, until, at the summer solstice, it is overhead at the Tropic of Cancer

($23\frac{1}{2}°$N). Now, as Fig 171 shows, it is the turn of the northern hemisphere to receive more than its fair share of daylight. Day is still 12 hours long at the equator, but at the latitude of London ($51\frac{1}{2}°$N) there are 16 hours of daylight out of the 24. It is also at this time that the noonday sun is highest in the sky. North of the Arctic Circle, the sun never sets. Even at midnight, the sun is still above the horizon, and these high latitudes enjoy 24 hours of daylight.

With the passing of the summer solstice, days in the northern hemisphere become progressively shorter. At the autumn equinox the sun is again overhead at the equator and day and night are everywhere of equal length. The hemisphere continues to experience shorter and shorter days until the earth returns to the winter solstice and the start of a new orbit.

experience sunrise at the time of the winter solstice, and their night lasts for 24 hours.

Gradually, day by day, the spinning earth swings round to the position of the spring equinox. Here, the tilt of the axis is neither towards nor away from the sun. Fig 170 represents this position, but we must imagine the sun's rays striking the page at right angles. The sun is vertically

Fig 169

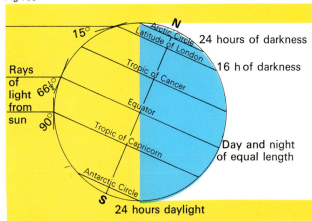

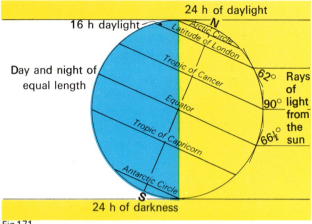

24 h of daylight

16 h daylight

Arctic Circle

Latitude of London

Tropic of Cancer

Day and night of equal length

62° Rays of light from the sun

Equator 90°

Tropic of Capricorn 66½°

Antarctic Circle

24 h of darkness

Fig 171

The seasons

A marked feature of life in British latitudes is the gradual transition from season to season. The buds of spring develop into the full summer foliage, which is shed in the autumn before the onset of the cold weather of winter. This progression of the seasons is another result of the tilt of the earth's axis. To understand it let us take another orbit around the sun.

The winter solstice, Fig 169, will again be our starting point. As we have seen, the noonday sun is overhead at Capricorn, and the northern hemisphere is tilted away from the sun. In British latitudes we experience the low temperatures of the winter season. This is partly explained by the shortness of the days, but a more important reason is the low angle of the sun. At London, the sun rises only 15° above the horizon. As will be explained fully in Chapter 13, when the sun's rays strike the earth at such an oblique angle, they have little power to raise temperatures.

Gradually the earth swings round to its other extreme. The vertical noonday sun, having passed the equator with the spring equinox, is overhead at Cancer at the summer solstice, Fig 171. This is summer. The days are longer and the sun is higher in the sky. Its elevation at London at the Solstice is 62°. Striking the earth at this steep angle, temperatures are high.

As the earth continues its orbit the days become progressively shorter and the sun's rays more and more oblique, until the position of the winter solstice is achieved once more. The fact that three days after the December solstice, many Australians spend Christmas Day basking on sun-drenched beaches is a reminder that the seasons are reversed in the southern hemisphere.

Finally, look again at Figs 169 and 171. Note that in low latitudes, days vary little in length and the noon day sun is always high in the sky. On the equator there is little difference between December and June. There is a monotony about the year in equatorial regions which is in marked contrast to the pronounced seasonal rhythm Britain knows so well.

12
The
Atmosphere

The atmosphere, which is such a vital part of man's environment, is the layer of gases which envelops the earth. The pie graph in Fig 172 shows that dry air is a mixture of several gases, the most important of which are nitrogen and life-giving oxygen. This mixture of gases extends outwards from the earth's surface for thousands of kilometres. With increasing altitude, however, the atmosphere becomes more and more rarefied, so much so that above a height of 50 kilometres, the amount of air is so small that it may be ignored. Fig 172 illustrates this, and also shows that of all the air in the atmosphere, over 50% is concentrated in a layer no more than 5·5 km deep. Note also that the higher parts of the earth's surface extend to altitudes where the air is much rarefied. Mountain climbers scaling the highest peaks must take supplies of oxygen with them. The Indians of the Bolivian Plateau, who live permanently at heights of over 4000 metres, are adapted to the thin air by having larger than average lungs, and a greater blood supply.

contain depends upon its temperature. The higher the temperature, the more water vapour it can hold. When air is holding its maximum amount of water vapour it is said to be *saturated*. Usually, air is less than saturated, and the amount of water vapour it contains, expressed as a percentage of the amount possible, is termed *relative humidity*. If, for instance, a volume of air at a certain temperature is holding only half as much water vapour as is possible, it has a relative humidity of 50%. Saturated air has a relative humidity of 100%.

Water vapour is invisible but we are frequently conscious of its presence. The humidity of the air influences our tolerance of temperature. Cold, dry air is preferred to the cold, 'raw' conditions when humidity is high. The 'muggy' combination of high temperatures and high humidity finds us reluctant to indulge in any activity, mental or physical. Under these conditions, even the most interesting lesson is hard to endure.

Although we are not aware of it, the atmosphere has weight and so exerts pressure on the earth's surface. The Italian scientist Torricelli, in 1643, was the first to demonstrate this atmospheric pressure with the famous experiment illustrated in Fig 173. The mercury in the glass tube does not flow down because it is supported by the pressure of the air on the mercury in the bowl. Imagine a column of air 1 cm² extending from sea-level to the edge of the atmosphere. The weight of the air

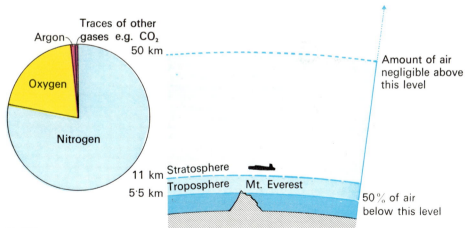

Fig 172

Fig 173

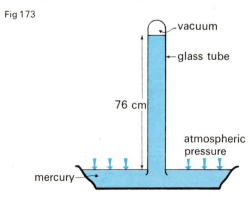

There is another gas always present in the atmosphere, though in small and variable amounts. This is water vapour, a dry gas not to be confused with the liquid water that may be present in such varied forms as cloud, rain or fog. The amount of water vapour the air may

in this column under average conditions would be 1033·3 grammes. Thus we say that atmospheric pressure at sea-level is 1033·3 grammes per square centimetre. At places above sea-level, it will be less, mainly because of the decreasing density of the atmosphere with increased altitude.

If we were to repeat Torricelli's experiment we would observe that the level of mercury in the tube would not remain constant for long. Its rise and fall would indicate frequent changes in atmospheric pressure. Our bodies are not sensitive to these small changes of pressure, but they are sufficient to set air in the motion that we experience as wind. Other aspects of the atmosphere are subject to changes that we can readily appreciate, and which frequently influence our daily lives. A heavy snowfall, for instance, may reduce traffic to a crawl, and a dense fog often raises hopes of an early end to the school day. An untimely shower may ruin the annual sports day, or a sudden drop in temperature may leave third man shivering on the boundary. Doubtless you will be able to add further examples from your own experience. The atmosphere is never constant, never still. It is a mass of gas of ever-changing conditions.

The variations in the state of the atmosphere from time to time and from place to place are studied as weather and climate. These terms are frequently confused, and the confusion is understandable, for both are concerned with the same elements – pressure and winds, temperature and rainfall and so on. When we talk of *weather* we are referring to the actual state of the atmosphere at a particular moment, or over a short period of time. The examples given in the last paragraph illustrate weather conditions.

By *climate* we mean the state of the atmosphere that may reasonably be expected to occur at a place over a long period, usually a month or a season. It may be regarded as average weather. An example or two will illustrate the difference. The Sahara desert has a hot, dry climate, but may experience a day or two of cloudy, showery

Fig 174

weather sometime during the year. The climate of the Highlands of western Scotland, where, on average, over 1500 mm of rainfall is recorded each year, may justifiably be described as cloudy and wet, and yet bright, sunny weather is by no means unknown.

Recording the atmosphere

The atmosphere is primarily the sphere of study of the meteorologist. In order to understand its processes and predict its moods he must have a wealth of detailed information on its various aspects. This information he obtains from a variety of sources. Specially-equipped weather ships are stationed in the Atlantic. Sensitive instruments are lifted high into the atmosphere by balloon, and send valuable information back to earth by radio. Television cameras in satellites, such as the American Tiros series, provide pictures of significant cloud patterns. In spite of such modern developments, the bulk of the information needed by the meteorologist is still obtained by careful recording at land-based stations.

In principle the recording stations maintained by the Meteorological Office differ little from the humble school weather station. Its instruments may be a little more sophisticated, observations more regular and frequent, and, dare we say it, perhaps a little more accurate, but both are devoted to recording the constantly changing state of the atmosphere. Let us now examine how essential measurements are made of the more important elements of weather and climate.

Temperature and humidity measurements are obtained with the aid of thermometers. The thermometers are housed 1·25 m above ground level in a *Stevenson Screen*. This is basically a box with louvred sides. These louvres are important for they allow the free circulation of air and keep out the direct rays of the sun, and so ensure that only true air temperatures are recorded. Fig 174 shows the arrangement of the thermometers inside the Stevenson Screen. The higher of the two horizontal thermometers records the highest temperature experienced in a particular period – usually 24 hours. The one below it records the lowest temperature for the same period. From these readings we can readily obtain a figure for the mean (i.e. average) temperature of a particular day. The thermometers mounted vertically are known as dry and wet bulb thermometers respectively. There is nothing special about the former, and from it the actual temperature at the time of observation may be recorded. The bulb of the other thermometer is wrapped in a muslin bag kept moist by a wick that leads to a small container of distilled water. The temperatures recorded by the dry and wet bulb thermometers are only identical when the air is saturated (i.e. when relative humidity = 100%). At other times the wet bulb gives a lower reading, and

the difference is an indication of the relative humidity of the air. The exact figure is read off from special tables.

Fig 175

Rainfall is recorded by a rain-gauge such as the one illustrated in Fig 175. This is simply a funnel supported in a copper cylinder which is set in the earth to give it stability. The rain that falls into the funnel runs down into a container. The amount of rainfall is measured by simply pouring it into the special measuring cylinder shown in the photograph. This is marked off in such a way that the number of millimetres of rain that have fallen per unit area can be read off directly. The measuring cylinder tapers to a point so that readings can be taken to within one-tenth of a millimetre. Less than this is recorded as a 'trace' of rainfall. The rain-gauge also records other forms of precipitation. Snow often presents problems. A light snowfall collected in the gauge may easily be melted and measured, but in heavy snow it is often quite a problem even to find the rain-gauge. In these conditions the depth of snow is measured and a rough approximation can be obtained for, on average, 10 mm of fresh snow is equivalent to 1 mm of rain.

Wind speed and direction are both of great interest to the meteorologist. Accurate measurement of wind speed is

Fig 176

given by an anemometer, one type of which is illustrated by Fig 176. The instrument is mounted at a standard height of 10 metres. The cups are attached to a spindle and as the wind blows the cups round, the spindle rotates. The greater the wind force the faster the rotation and, on a modern instrument, the speed is shown on a dial in knots. A knot is a speed of 1 nautical mile (1·85 km) per hour. The anemometer is often linked to a sensitive wind vane which records wind direction.

If delicate and expensive instruments are not available, we can, by observation, obtain approximate values for wind speed and direction. The latter is shown, for instance, by the plume of smoke from a chimney, but remember that the direction of the wind is the direction it is coming from, not its destination. Wind speed can be gauged by its effect on such common features as trees. A light breeze of 5 knots is sufficient to cause the leaves to rustle, but for whole trees to be in motion, a moderate gale with wind speeds of about 35 knots is needed. These examples are taken from the Beaufort Scale which classifies wind speed into 13 categories from Calm to Hurricane. Look up this scale in a textbook of meteorology. With its help, and after a little practice, you will soon be able to make very reasonable estimates of the speed of the wind.

Pressure changes, as we saw above, are indicated by the variations in height of a column of mercury. Torricelli's simple glass tube was the forerunner of the modern highly-sensitive mercury barometer. Pressure may also be measured by an aneroid barometer. This instrument depends upon a small, air-tight drum made of thin, corrugated metal. It contains a partial vacuum, and when atmospheric pressure is high the drum is compressed, and when pressure falls, it springs back. This movement is emphasised by an arrangement of levers, and the changes are recorded on a dial. An aneroid barometer is a handy instrument, but it is slightly less accurate than the mercury barometer favoured by meteorologists. The unit of pressure is the *bar* which is divided into 1000 millibars (mb). Average pressure is 1013 millibars.

In recording pressures, corrections are made for the effect of altitude, and the figures usually quoted and mapped refer to pressures at sea-level.

Sunshine is recorded by the instrument shown in Fig 177. The sunshine recorder is adjusted for the latitude of the station, and the glass ball focuses the sun's rays on to a piece of sensitive card. When the sun is shining a burn mark is made on the card, and the length of the mark gives the duration of sunshine.

Visibility and Cloud conditions are recorded by observation. If the distance from the station to a number of prominent features is known, maximum visibility can be obtained by noting the furthest of these features that can be seen. The extent to which the sky is obscured by cloud is estimated by eye to the nearest eighth, and the type of cloud is also noted.

Fig 177

Climate statistics and maps

When the meteorologist makes his observations at, say, 09.00 hours, he is, of course, measuring weather conditions. To obtain statistics of climate, his readings must be averaged out over a long period of time – at least 30 years. For example, the average temperature of a particular month is obtained from the daily readings. Then the figures for this month in the preceding 30 years are averaged to give the mean monthly temperature. Mean annual temperatures are obtained by adding the monthly figures and dividing by 12. All the elements of climate may be averaged in this way and statistics are thus obtained which collectively give a good indication of climatic conditions.

The distribution of climatic elements is shown on maps by *isopleths*, i.e. 'lines of equal value'. The isopleth for rainfall is called an *isohyet*. It joins all places on a map that experience the same average rainfall either over the year or for a month or season. Similarly, pressure and temperature conditions are shown by *isobars* and *isotherms* respectively. Your atlas contains many examples of this type of climate map. If you examine one or two you will notice that the elements of climate show great regional variation. The following chapters will suggest reasons for these variations.

13

Temperature

The heat on which all forms of life depend, is derived from the sun. This star, with temperatures, even at its surface, of the order of 6000°C, radiates energy out into space in all directions. Only the tiniest fraction is destined for our small planet. Our share of solar radiation, or *insolation*, is constant, yet temperatures experienced on Earth vary greatly from place to place. These variations are the result of the factors considered below.

Altitude

Insolation, because of its very short wave length, has very little effect on the temperature of the air. It is, however, readily absorbed by the surface of the earth, which re-radiates the heat thus gained back into the atmosphere. This *terrestrial radiation* is of a long wave length, and readily raises the temperature of the air, especially in the lower layers of the atmosphere where water vapour and dust are most abundant. Also, air in contact with the earth's surface is heated by conduction and circulated upwards by air movements. Thus, perhaps surprisingly, the atmosphere is not heated directly by the sun, but from the earth. Were we to ascend through the atmosphere in a balloon we would be moving away from the source of heat, and with a thermometer we could record this decrease of temperature with altitude. On average it would be 0·6°C per 100 metres.

We do not need to have a balloon to observe these temperature changes. They can readily be appreciated by an ascent even of one of Britain's modest mountains. A climb (or, for the less energetic, a short rail journey) from Llanberis to the summit of Snowdon means a fall in temperature, on average, of about 6°C. On a sunny day, the rocks at the summit may well feel warm, but the clear air presents little resistance to terrestrial radiation and the air remains cool.

This reduction of temperature with altitude affects man in many ways. Quito, the capital of Ecuador, is within a degree of the equator, but its average annual temperature is only 13°C. This is because it lies in an Andean valley

2850 metres above sea-level. Towns in the Himalayas, such as Simla at a height of 2300 metres, are resorted to by people wishing to escape from the stifling heat of the plains just 60 kilometres to the south. Plant life is sensitive to these changes of temperature. Natural vegetation and the crops man may cultivate show sharp changes with altitude. High mountains in low latitudes may show a complete transition from tropical conditions at sea-level to permanent snow and ice near the summit.

Latitude

On the surface of the earth there is a gradual reduction of temperatures with increasing latitude. The contrast between the hot tropics and the permanent snow and ice of polar regions is an obvious illustration. Fig 178 helps to explain this variation. A and B represent equal bands of insolation, and X and Y the areas of the earth's surface that they cover. Because A strikes the earth at a more oblique angle, X is larger than Y. The heating power of A being spread over a greater area is, therefore, less effective and results in lower temperatures. Try the little experiment illustrated in Fig 179. The more obliquely the torch shines on the book the larger, but less bright,

Fig 178

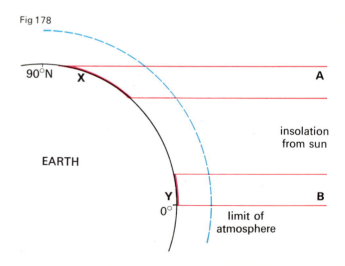

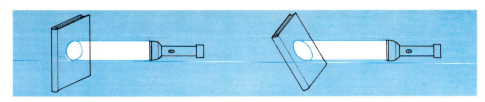

Fig 179

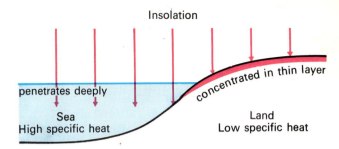

Fig 181

is the area illuminated. With insolation instead of light the same is true of the earth. From Equator to Pole, insolation becomes more oblique and therefore less effective, hence the gradual lowering of temperatures towards the Poles.

Another factor increases this effect. To reach the earth's surface, insolation must pass through the atmosphere. In doing so, much is reflected back into space by clouds and dust particles. The greater the distance through the atmosphere, the greater will be the loss. Fig 178 shows that A has the longer journey and thus suffers a greater reduction than B.

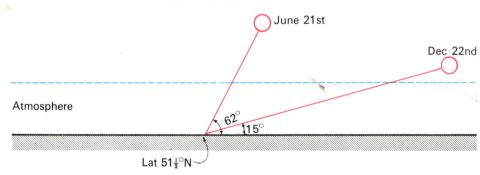

Fig 180

Incidentally, these same effects explain seasonal and daily variations in temperature. Fig 180 shows, for the latitude of London, the angle of the midday sun at the solstices. Note how obliquely the winter insolation reaches the earth's surface. The seasonal contrast is heightened by the greater length of the summer day. Similarly, the low sun at dawn has great difficulty in taking the chill off the morning, but as the sun's rays become less oblique, insolation becomes more effective and the temperature rises.

The effect of land and sea

The position of a place in relation to the distribution of land and sea has a great influence on the temperature it experiences. The reason for this is that insolation varies in its effect on these two types of surface. Fig 181 shows that insolation striking the land is absorbed by a very thin layer. Over the sea, however, because insolation penetrates deeply, its heating power is spread through a much greater volume. More important than this is the fact that the sea has a higher specific heat. This means that water must absorb more insolation than land for temperatures to be raised by the same amount. Or, to put it another way, with equal amounts of insolation the temperature of the land will rise much more than that of the sea. On a sunny summer's day the land is much hotter than the sea. Perhaps you have observed this on a seaside holiday. The bather who runs happily into the sea frequently beats a shocked and shivering retreat to the warmth of the sandy beach.

Another difference between land and sea is that the former loses heat by terrestrial radiation much more readily. It is therefore quick to warm up and as quick to cool down. Land surfaces show great ranges of temperature, both diurnally and seasonally. The opposite is true of the sea; it shows little variation in temperature. Winter bathers are perhaps less brave than we imagine: the chief test of their courage is not so much entering the water as crossing the beach.

As the atmosphere is heated by the earth's surface, air temperatures reflect these differences between land and sea. Air over the land masses shows a marked contrast in temperature between summer and winter but, over the oceans, air temperature shows only a small annual range. This gives rise to the important distinction between 'continental' and 'maritime' climates, the influence of which is greatly extended by the next factor to be considered.

Winds and ocean currents

The influence of wind and currents on temperatures is well illustrated in that part of the world covered by Fig 182. This map shows the direction of the prevailing winds in these latitudes. It is a variable air stream, but is predominantly south-westerly in direction. Thus the air it brings to Britain has its origin over the ocean. It is maritime air – relatively warm in winter and cool in summer, and these characteristics are reflected in temperatures recorded in Britain. In contrast, the south-westerly winds transport air of continental origin to the eastern coast of North America, and this area experiences hot summers, cold winters and a great annual range of temperatures.

The temperature difference between the western and eastern shores of the north Atlantic is most marked in winter. Fig 182 shows that the St. Lawrence is regularly closed to shipping by winter ice, a hazard unknown on the Mersey, which is over 6 degrees nearer the Pole. Indeed, ice clutters the western side of the ocean basin as far south as Newfoundland, whereas, on the east, ice-free seas extend well beyond the Arctic Circle. You will note how the January isotherm of 0°C bulges northwards. This so called "gulf of winter warmth" is due principally to the prevailing south-westerly winds, but ocean currents play a supporting role. The Gulf Stream, with its extension the North Atlantic Drift, originates in warm sub-tropical waters, and transfers heat to northern latitudes. It also warms the winds passing over it, and so raises air temperatures. An ocean current which has the opposite effect is the Labrador Current which moves cold polar waters, studded with icebergs, to southerly latitudes.

Maritime influences become less pronounced with increasing distance from the sea. Europe and western Asia shows a gradual transition between maritime and continental climates, and Fig 183 illustrates this diagrammatically. July and January average temperatures are given for a number of stations all on or near latitude $51\frac{1}{2}°$N. Note that towards the interior of the land mass summers are warmer and winters colder, with a corresponding increase in the annual range of temperature. Even within the British Isles, increasing distance from the Atlantic produces significant differences in temperature.

The factors discussed above account, in broad outline, for the variations of temperatures over the surface of the earth. In detail, however, other factors account for variations of temperatures that are of local importance and thus merit brief consideration.

Fig 182

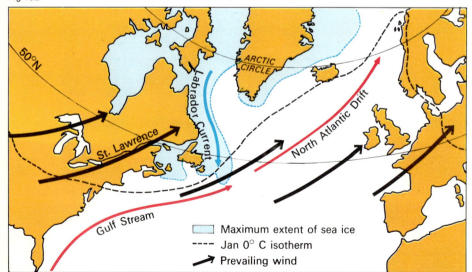

50°N

ARCTIC CIRCLE

Labrador Current

St. Lawrence

North Atlantic Drift

Gulf Stream

Maximum extent of sea ice
Jan 0° C isotherm
Prevailing wind

Temperature

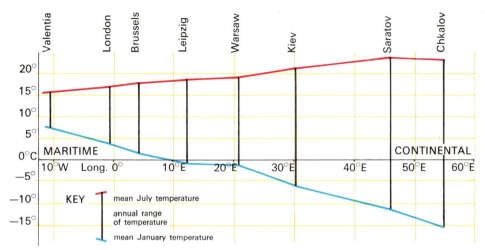

MARITIME ... CONTINENTAL

KEY

— mean July temperature

— annual range of temperature

— mean January temperature

Fig 183

Clouds

As we have seen, clouds act as a barrier to insolation. Thus the world's highest temperatures are recorded not in cloudy equatorial regions, but under the clear, sunny skies of tropical deserts. Clouds also check terrestrial radiation, and cloudy areas tend to have a small diurnal range of temperature. Britain's lowest temperatures are recorded on winter nights with clear starlit skies, when there are no clouds to check the loss of heat.

Mountain barriers

High ranges of mountains impede the circulation in the lower atmosphere. The high Himalayas, for example,

Fig 184

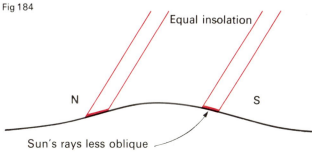

Equal insolation

N   S

Sun's rays less oblique

protect northern India from the cold winter air of interior Asia. The Rockies restrict the penetration of maritime air from the Pacific and as a result the interior of the North American continent has more pronounced continental temperature conditions.

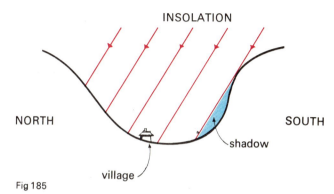

INSOLATION

NORTH   SOUTH

shadow

village

Fig 185

Relief

The relief of the land surface results in many temperature variations which, though small, are often of significance to man. South-facing slopes (in the Northern Hemisphere) are favoured for vine cultivation for reasons Fig 184 will make clear. Similarly, Fig 185 accounts for the location of Alpine villages on the north side of east-west valleys. The location of orchards provides another

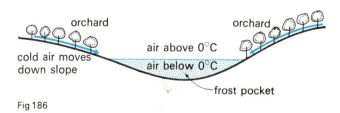

orchard   orchard

air above 0°C

cold air moves down slope   air below 0°C

frost pocket

Fig 186

example. Fig 186 shows that trees are planted on the sides of the valley. This is because heavy, cold air moves down the slopes and a 'frost pocket' of air below freezing point may accumulate in the valley bottom.

14 Pressure and Winds

Chapter 14

Winds

The unequal heating of the atmosphere described in the last chapter leads to great variation in atmospheric pressure. This is because, as air is heated it expands, becomes less dense and therefore exerts a lower pressure. In general terms the warmer the air the lower the pressure. Wind is the movement of air from high to low pressure.

Fig 187 illustrates a difference of pressure as it would be shown on a weather map. Note that pressure changes gradually from high to low. The isobars indicate a *pressure gradient* just as contours reveal the gradient of the land surface. The closer the isobars the steeper the gradient. Wind speed is related to this gradient – the steeper the gradient, the greater will be the wind speed. Isobars bunched together indicate strong winds, whereas a weather map with few isobars indicates little difference in pressure and hence gentle breezes or calm air.

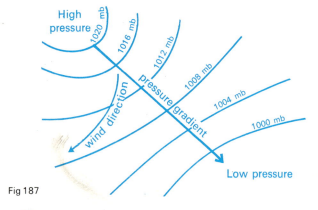

Fig 187

Unlike water flowing down a slope, air does not move in a straight line from high to low. In Fig 187, wind direction is seen to cut the isobars at an angle. Moving air is greatly influenced by the fact that it is moving over an earth that is rotating on its axis. This rotation causes winds to be deflected to the right in the northern hemisphere and to the left in the southern.

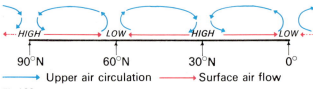

Fig 188

Planetary winds

Although atmospheric pressure shows great variation, it is possible to identify major belts or zones where the pressure is, on average, relatively high or low. Fig 188, which represents the surface of the earth from equator to North Pole, shows the approximate position of these belts. Low pressure is seen to be centred on the equator and latitude 60°N and high pressure is characteristic of the polar region and 30°N. The low pressure at the equator is the result of intense heating, and the low temperature of the polar region produces high pressure. The explanation of the other belts, high at 30°N and low at 60°N, demands a knowledge of meteorology beyond the scope of this book. Air moves from high to low pressure as major or *planetary* winds. The atmospheric circulation is completed at high levels as shown in the diagram, and this pattern of air movement is repeated in the southern hemisphere.

Fig 189 shows in a very generalised way, the distribution of pressure belts and planetary winds over the surface of the earth. This simple pattern is, in reality, much modified by factors to be considered later. Note that the direction of the winds shows the influence of the earth's rotation. The names given on the diagram have a nautical flavour and knowledge of the wind system was essential in the days of sailing ships. Before the days of steam, ships took a southerly course across the Atlantic to benefit, as Columbus had first done, from the moderate but reliable North-East Trades. They sailed home with the variable and often stormy south-west winds of more northern latitudes. On voyages to the southern hemisphere the Doldrums were feared as an area of calm air

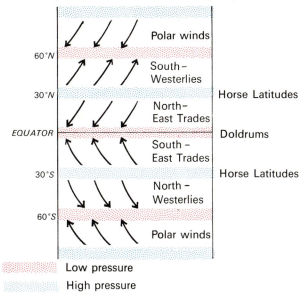

Fig 189

Low pressure

High pressure

that frequently led to vessels being becalmed for days on end. The wool clippers making the round voyage from London to Australia and back, left the Atlantic via the Cape of Good Hope and returned by rounding Cape Horn, thus to make best use of the strong north-westerlies known as the Brave West Winds. It is interesting to note how closely the route of the round-the-world yachtsman of today corresponds to that of the old clippers.

Shift of the wind system

The Doldrum belt of low pressure, the driving force of the world wind system, is shown in the Fig 189 as being centred on the equator. This, however, is only its average position. Being caused by the intense insolation of the overhead sun, its position reflects the apparent movement of the overhead sun discussed in Chapter 12. In June, for example, when the sun is overhead in the northern hemisphere, the centre of the Doldrum belt is also north of the equator. The other pressure

belts, and hence the planetary winds, follow the lead of the Doldrums in the movement known as the 'shift of the wind system'. This shift lags behind the overhead sun, and covers only about 10° of latitude.

Fig 190 shows the shift of the wind system diagrammatically. In (i) the sun is overhead at the Equator and the Doldrums are centred on that latitude. With the movement of the overhead sun to the Tropic of Cancer (ii), the Doldrum belt is dragged northwards, only to return to the equator at the September equinox (iii). The December solstice (iv) sees the Doldrums extended into the southern hemisphere. Note from the diagram how the high pressure of the Horse Latitudes and the planetary winds show a similar movement.

Fig 190

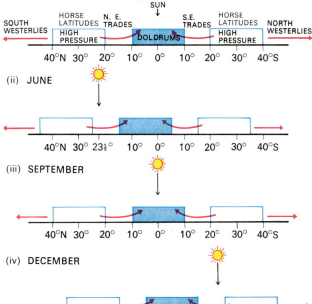

The shift of the wind system, although of comparatively small extent, has a very important effect on world climates. This can be appreciated from Fig 190. The Doldrum belt is associated with high temperatures and humidity and heavy rainfall, whereas land areas of high pressure and Trade Winds are characteristically dry. The diagram shows that only a belt between 5°N and 5°S experiences Doldrum conditions all the year round. Polewards of these limits, for part of the year, Doldrums are replaced by the dry Trade Winds, giving a pronounced seasonal contrast of climate. Another seasonal contrast is indicated on the diagram between latitudes 35° and 40°, where rain-bearing Westerlies alternate with dry high pressure conditions.

Monsoon winds

Before the advent of the steamship, trade in the Indian Ocean was dominated by Arabian sailors in their single-masted dhows. Trade was very seasonal. The voyage from Arabia to the coast of India was only possible in summer when south-west winds were dominant. To return to Arabia, the dhows had to await the north-east winds that prevailed in winter. From the Arabic word for 'season' is derived our word *monsoon* which means a large scale seasonal reversal of wind direction.

Monsoon winds are the result of the unequal heating of land and sea. Fig 191 shows the areas where monsoon effects are most pronounced. In winter the land mass of Asia loses heat rapidly by terrestrial radiation. Intense cold leads to very high pressure and cold, dry air moves outwards from this high pressure in the directions shown. In summer the land heats up rapidly and low pressure is the result. At this season warm, moist winds sweep in from the sea as the summer monsoon. The monsoonal effect completely masks the normal planetary wind system shown in Fig 189.

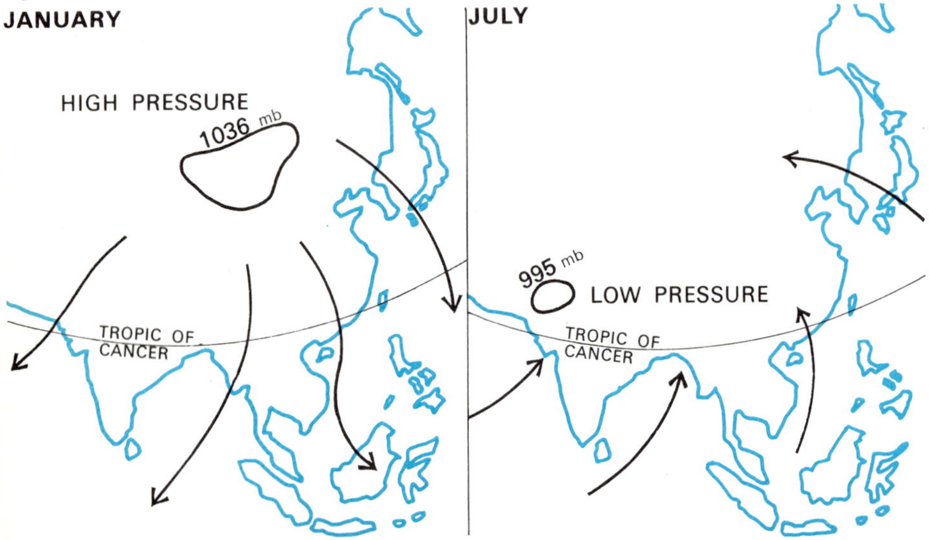

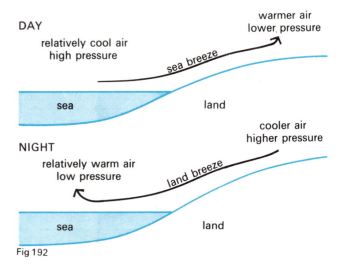

Fig 192

Land and sea breezes

On a hot summer's day at the seaside we are often thankful for a cool breeze moving in off the sea which moderates the heat. Were we to remain on the beach after nightfall we would probably notice this sea breeze replaced by a breeze blowing off the land. These local winds have the same basic cause as Asia's mighty monsoons – the unequal heating of land and sea.

Fig 191

JANUARY

JULY

HIGH PRESSURE

1036 mb

TROPIC OF CANCER

995 mb LOW PRESSURE

TROPIC OF CANCER

Examine Fig 192. During the day, air over the sea is heated less than the air over the land. The higher pressure therefore, is over the sea and air moves inland as a sea breeze. At night, however, the more rapid cooling of the land surface causes pressure to be relatively high over the land and air moves from land to sea.

Hurricanes and tornadoes

The full power and fury of air in motion is revealed in these violent storms. The centre or 'eye' of a *hurricane* is between 15 km and 30 km across. Here atmospheric pressure is very low and the air relatively calm. Around the eye, however, is a zone over 300 km wide, where the pressure gradient is steep and winds blow anti-clockwise at speeds often in excess of 160 km/h. Hurricanes originate over the warm waters of the Atlantic between 10° and 15°N. They travel north-west at speeds of about 15 km/h and swing round to the north-east before they slowly fade away. Their track often takes them over the Caribbean Islands and the mainland of North America. When this happens high winds and torrential rain cause great damage to crops and property. The havoc is increased in coastal areas by flooding due to exceptionally high tides.

Hurricanes occur mainly in the summer months. Each individual storm, identified by a girl's name, is closely tracked by aircraft and radar so that adequate warning of its approach may be given. Storms of this type are known as typhoons in the China Sea, cyclones in the Bay of Bengal and willy-willies off the north-west coast of Australia. They are also unwelcome visitors to Queensland and the Malagasy Republic.

The world's most destructive storms are the *tornadoes* which are particularly frequent in the Mississippi Valley of the U.S.A. They are usually less than 500 metres in diameter and seldom last more than an hour or two, but they leave a path of total destruction in their wake. Pressure is so low in the eye of a tornado that houses explode outwards. More damage is done by the extremely high wind velocities. It is impossible to give a figure for wind speed, for anemometers are included in the general destruction, but it is believed to be in excess of 800 km/h. Similar storms over the oceans give rise to *waterspouts*. As a result of the extremely low pressure, a solid column of water may rise more than 3 metres above the surface of the ocean and spray is carried high into the air.

15

Water in the Atmosphere

Even on the hottest summer's day there is water in the atmosphere. We may not be able to see it, but it is always present in the form of water vapour. The significance of this dry gas is far greater than its modest share of the atmosphere – always less than 5% – would suggest. From it is derived the water in liquid or solid forms that so frequently intrudes into our daily lives and which forms such important elements of weather and climate Cloud, rain, fog, dew, snow and hail are all the result of a single basic process – the condensation of water vapour.

Fig 193 will help us to understand this important process. The graph shows the amount of water vapour contained in saturated air at different temperatures. It emphasises the point made in Chapter 12, that the amount of water vapour air can hold depends upon temperature. The graph's steeply-rising curve shows that with higher temperatures the ability of air to hold water vapour increases dramatically. Read off the values for air at 35°C and 25°C It will be seen that air at the higher temperature can hold almost twice as much water vapour.

Fig 193

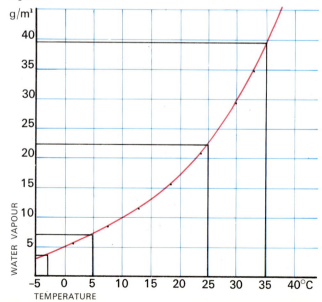

Imagine a cubic metre of air at 35°C which contains 22 grammes of water vapour. A glance at the graph will show that this air is far from saturated. In fact, it only contains about half the possible amount of water vapour. But if this air is cooled to 25°C, then it will be holding its maximum – it will be saturated or at *dew point*. If the air is cooled below 25°C its ability to hold water vapour is further reduced and the surplus vapour is condensed and appears in the atmosphere as tiny droplets of liquid water.

This process is easily illustrated from our own experience. Think of a classroom on a cold winter's day. As we breathe out and perspire we add moisture to the air and humidity rapidly builds up. The air in contact with the cold window panes is cooled to and below dew point and the water vapour that can no longer be held in the air is condensed into drops of water that cover the glass. The windows, as we say, become 'steamed up' and provide a tempting surface for artistic fingers.

Let us return to Fig 193 for another example. This time our cubic metre of air is at 5°C and contains 3·5 grammes of water vapour. From the graph we see that for this air to be saturated it must be cooled to −3°C. Any further reduction of temperature leads to condensation, but because it takes place below freezing point, it is not drops of water that result, but tiny crystals of ice.

It is, then, the cooling of air below saturation point that is responsible for the occurrence in the atmosphere of water in all its varied liquid and solid forms. There are two main ways in which this cooling may be brought about. We have already had an example of one way – cooling by contact. When the surface of the earth is cold, it acts like the window pane in the classroom and cools the thin layer of air in contact with it. With sufficient cooling the air temperature will fall below dew point, and condensation will result in the formation of dew, fog or frost depending on circumstances.

The second process is cooling by expansion. Air is a gas

WINDWARD

LEEWARD

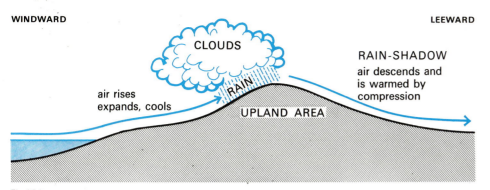

CLOUDS

RAIN-SHADOW
air descends and
is warmed by
compression

air rises
expands, cools

RAIN

UPLAND AREA

Fig 194

Types of rainfall

For air to cool by expansion it must rise in the atmosphere. There are three ways in which this ascent may be brought about and each gives rise to a different type of rainfall.

a. *Relief (orographic) rainfall*

Examine Fig 194 on which the arrow indicates the movement of air. As it blows inland it meets the upland and is forced to rise. As it rises it moves into areas of lower pressure and therefore it expands. This expansion leads to cooling which is often sufficient to cause condensation and the formation of cloud and rain. On the other side of the upland, the air descends into areas of higher pressure and is warmed by compression. There is no question of cooling and condensation. The diagram shows that rain is confined to the windward side. The leeward side forms what is known as a *rain-shadow* area.

When relief barriers of great altitude are exposed to prevailing winds from the sea, this rain-shadow effect can lead to dramatic contrasts in rainfall. The Andes, for instance, separate southern Chile, which is deluged by over 2500 mm annually, from Patagonia which counts itself lucky if it receives 200 mm. The rainfall maps in your atlas will provide further examples. Have a look at South Island, New Zealand, for instance. Even in Britain where upland areas are of modest elevation the rain-shadow effect produces significant contrasts. The sheltered Vale of Eden, for example, receives less than half the amount of rain that falls on the uplands of the Lake District, which lie only a short distance to the west.

Look again at Fig 194, for it helps to explain a local wind of considerable significance in mountainous districts. As rain is formed, the heat known in physics as the latent heat of condensation is given out and absorbed by the air. This means that the air is warmer at the foot of the leeward slope than it was before it started to rise. In high mountains, this rise of temperature may be

and faithfully obeys the laws that govern the behaviour of all gases. One of these laws states that when a gas is compressed its temperature is increased. We are made aware of this when we inflate a bicycle tyre, for the hand holding the pump feels the warmth produced by the compression of air inside the pump. The opposite is also true: when a gas expands it cools. When air rises through the atmosphere it is moving into layers of lower pressure and hence it will expand and cool.

Provided air rises sufficiently, this cooling by expansion readily leads to the condensation of water vapour and the formation of tiny droplets of liquid water. These droplets are too small to fall through the atmosphere. They are suspended in the air in one of the many types of cloud. Inside a cloud, droplets are frequently in collision and as a result they increase in size until they are large enough to overcome the resistance of the air and fall as rain. If dew point is below freezing point, condensation will lead, of course, to the formation of tiny crystals of ice. At higher levels, say, 6000 metres or so, clouds are always composed of ice crystals. They are held in suspension until they amalgamate into snowflakes that are large enough to fall through the atmosphere. Many of these snowflakes are destined never to reach the earth, for they often melt in the higher temperatures of lower levels and are recorded as rain or sleet.

considerable. It is especially significant in the deep troughs of Alpine valleys where the descending air is known as a *föhn* wind. This warm, dry wind is a mixed blessing. In late spring it rapidly melts the winter snow and brings an early start to farming activities. At other times it can lead to disastrous avalanches. A similar wind descending over the Canadian Rockies is called the *chinook*.

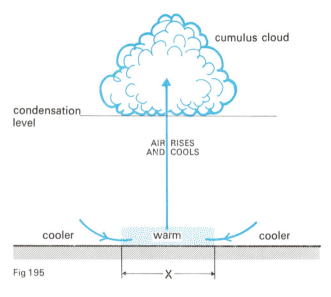

Fig 195

b. *Convectional rainfall*

This type of rainfall is explained by Fig 195. It is assumed that, as a result of unequal heating of the earth's surface, area X is raised to a higher temperature than the earth on either side. The air above X becomes warmer than the surrounding air. Being warmer it is less dense and will rise through the atmosphere by convection. Ascent brings cooling by expansion and if condensation level is reached, billowing cumulus clouds will form, Fig 196. Because of rising air currents, droplets must grow to a large size before they can fall, hence convectional rain typically occurs in heavy showers of short duration. The rainfall of equatorial areas is predominately of this type.

Fig 196

Thunderstorms result from pronounced convectional currents in the atmosphere. Rapidly rising air currents lead to the formation of the towering cumulonimbus type of cloud with the characteristic 'anvil' shape illustrated by Fig 197. This thunder cloud may be only a few kilometres across, yet have a vertical extent of thousands of metres. Within it, due to processes not yet fully understood by scientists, electrical charges accumulate in the pattern indicated in the diagram. Voltage differences build up to something like 100 million volts, and their discharge gives us the giant electrical sparks we know as lightning, Fig 198. The discharge is usually

Fig 197

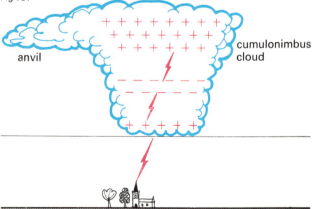

Fig 198

It is the type of rain associated with the depressions which are such frequent, and often unwelcome, visitors to our shores. It will be treated in detail in Chapter 17, and for the present, Fig 199 will suffice. Two currents of air at different temperatures come together. The lighter, warmer air rises over the cold air and cooling by expansion leads to precipitation.

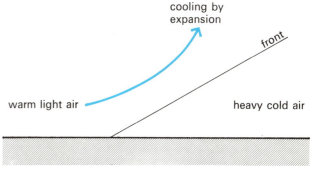

Fig 199

between different parts of the cloud but sometimes lightning sparks from the cloud to the earth. It is attracted to the higher parts of the earth's surface, hence the lightning conductors on tall buildings and the danger of sheltering under a tree in a thunderstorm. Thunder is simply the noise created by lightning as it disturbs the air in its path.

The thundercloud is so tall that its upper levels are below freezing point. The rising air currents which sustain it are variable in strength. A raindrop, on a strong updraught, may be carried to higher levels and so freeze, only to fall again, as the current slackens, to lower levels. Here the frozen raindrop collects more water which freezes on the next upward thrust. This process, repeated many times, produces spherical pellets of ice that grow until they are large enough to fall as *hail*. In Britain, hailstones are usually of modest size, but in warmer climates, hailstones the size of golf balls, or even cricket balls, are not uncommon. Hailstones of this size do considerable damage to standing crops. The largest recorded hailstone, incidentally, was one which fell in Nebraska, U.S.A., in 1928. This monster was 430 mm in circumference and weighed 680 grammes, no less.

c. *Cyclonic (frontal) rain*

This is the most important type of rain in our latitudes.

Other forms of precipitation

Fog consists of tiny droplets of water suspended in the atmosphere. In this it resembles cloud, but fog occurs at ground level. Its chief significance lies in the reduction of visibility that it causes. Even a moderately dense fog plays havoc with transport services. Aircraft are diverted, trains cancelled and road vehicles play 'follow my leader'. In a real 'pea souper' even the pedestrian finds progress difficult.

The fog most commonly experienced over land areas is known as *radiation fog*. It occurs most frequently on clear nights in the winter months. Under clear skies there is no check to terrestrial radiation and the temperature of the earth's surface rapidly falls. The air is cooled by contact with this cold surface, and with a slight air movement, this cooling is extended to a considerable depth of the atmosphere. When the air temperature falls below dew point, condensation produces water droplets which remain in suspension.

Fig 200

Particles of solid matter, especially those produced by the burning of coal, are trapped in the lower atmosphere when fog is present. The resultant mixture is known as *smog*. In persistent fogs solid material may accumulate to such an extent that it presents a risk to health, especially for people suffering from respiratory ailments. The fog that blanketed London in December 1952 is estimated to have cost the lives of over 4000 people. With the recent development of a clean air policy involving the creation of smokeless zones, the incidence of dangerous smog has been much reduced.

Although radiation fog is restricted to land areas, the sailor does not escape the hazard, for other factors lead to the formation of fog over the sea. The waters off the coast of Newfoundland are notorious for the frequent occurrence of fog. In this area, as Fig 182 shows, warm and cold currents meet. Warm, humid air moving from over the Gulf Stream is readily cooled when it meets the icy waters of the Labrador Current, and so fog is formed. The seas north of Japan are similarly afflicted and for the same reason.

Mist differs from fog only in the degree to which it restricts visibility. One kilometre is the dividing line. If visibility exceeds this figure we have mist rather than fog. Early morning mist over Elterwater is shown in Fig 200.

Dew is formed under the same conditions as fog, except that the air is not in motion. With still air, cooling is restricted to the lowest layer of the atmosphere. As the temperature of this layer falls below dew point condensation occurs and drops of water are deposited directly on the surface of the earth. Vegetation is particularly favoured for, due to moisture given out by the plant, the air in contact with the leaves is of a high humidity and little cooling is needed to produce condensation.

Hoar frost is formed in the same way as dew except that dew point is not reached until the temperature is below 0°C and therefore crystals of ice are deposited as in Fig 201.

Fig 201

16
The
Weather
Map

In Chapter 12 we saw how recordings and observations of the atmosphere are made. The information thus gained is coded for brevity and sent as speedily as possible to the Meteorological Office at Bracknell, Berkshire. Here it is joined by similar information from recording stations in Europe and North America, from Atlantic weather ships and from balloons and aircraft. In all, over 2 million observations are received at Bracknell each day. A selection of this wealth of detailed information is plotted on special maps known as *synoptic charts* or, more popularly, as *weather maps*. The synoptic chart gives a clear and precise picture of the state of the atmosphere at a particular moment in time, and is the basis on which the weather forecaster makes his predictions.

A simplified version of the forecaster's chart is included in the Daily Weather Report issued by the Meteorological Office. The maps on page 100 are taken from this publication, which is available by post to the general public. When you turn to these maps do not be discouraged by their apparent complexity. With the aid of the key, Fig 202 and a little patient study, their secrets will be unlocked and they will soon become as meaningful as more familiar types of maps.

The site of each recording station is represented by a small circle. This circle is also used to show the amount of cloud at the time of observation. The observer estimates the proportion of the sky obscured by cloud in *oktas* or eighths, and this information is shown by the scale of shading given in Fig 202. The direction of the wind is that of the small line drawn to the station circle. The dashes added to this line give an indication of the wind speed in knots. A short dash represents 5 knots and each larger one has twice this value. A solid triangle is used when wind speed is in excess of 48 knots. The air temperature at each station is given in degrees Celsius, and the nature and the type of precipitation is shown by symbols. To the weather forecaster, the most significant information he records on his synoptic chart is that concerning

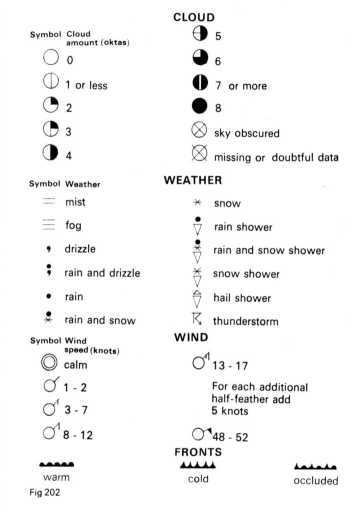

Fig 202

atmospheric pressure. This is shown by isobars drawn at intervals of 4 millibars. From the maps on page 100 we can appreciate the complexity of the pressure distribution thus recorded. It is usually possible, however, to identify certain distinctive patterns in the isobars. These are significant because each is associated with a particular set of weather conditions.

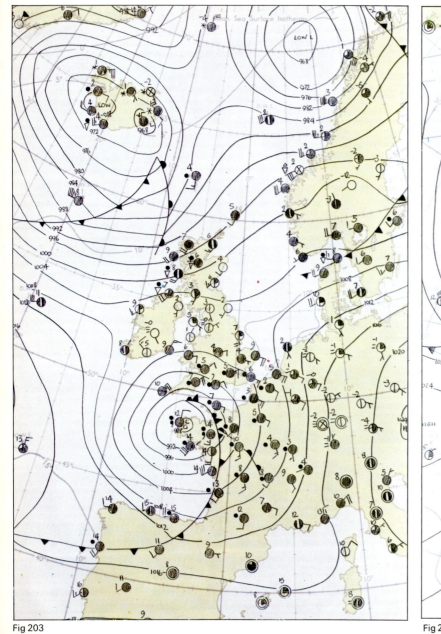

Fig 203

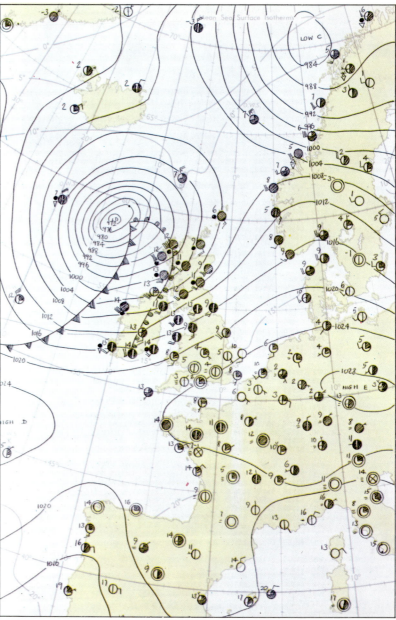

Fig 204

One of these important patterns is well shown on Fig 203. Note the set of closed isobars, with the lowest pressure in the middle, that is centred over north-west France. This is a *depression* or *low*. The symbols for the various fronts, Fig 202, have been drawn in to show where masses of air with different characteristics come together. Another example of a depression is shown on Fig 204 to the north-west of the British Isles.

An *anticyclone* or *high* may be recognised by widely spaced, closed isobars with the highest pressure in the middle. On Fig 204, the isobar for 1028 millibars is the centre of an anticyclone that extends well beyond the eastern limit of the map. It is labelled 'HIGH E' to distinguish it from a similar feature over the Atlantic.

There are other patterns the meteorologist looks for among the isobars on his synoptic chart. Col, ridge and trough, though beyond the scope of this book, are three examples you will soon come across if you follow the weather forecasts on television or in the newspapers. These, like depressions and anticyclones, are distinctive arrangements of pressure, each associated with a particular weather situation. They form part of a constantly changing atmosphere and are seldom stationary for long. They move in procession over the British Isles bringing frequent changes of weather. Their movement is fickle: it is subject to subtle changes of speed and direction that sometimes frustrate the prediction of even the most skilful weather forecaster. These travelling pressure systems are largely to blame for the notorious variability of British weather, which is the subject of the following chapter.

The best way to become familiar with weather maps is to use them. The two examples in this book will give you a good start. Select a station at random and, with the aid of the key, work out the weather situation as it was recorded by the meteorologist. Let us take a couple of examples from Fig 204. In the Shetland Isles where the barometer stood at 1004 millibars, rain was falling from a completely overcast sky. Air temperature was 6°C and a wind of 13 to 17 knots was blowing from the south-east. East Anglia had a much brighter start to the day. A temperature of 10°C was recorded and pressure was 20 millibars higher than in Shetland. There was not a trace of cloud in the sky, and the speed of the southerly wind was 8 to 12 knots.

With a little experience you will notice significant contrasts in weather between different areas. On Fig 203, for instance, work out the different weather conditions experienced in southern and in northern England, and between the north and south coasts of Iceland. On this map north-west France and central Europe may also be contrasted. On Fig 204 it will prove profitable to compare the weather over north-west Scotland and central Europe.

Another way in which you will become familiar with the weather symbols is to use them to record your own weather observations. At the moment of writing there is not a glimmer of blue in a low overcast sky. The rain, driven by a south-west wind of about 40 knots, is lashing against the windows. The thermometer records a temperature of 8°C. Using symbols this could be written ⦚⦙●. It is the same frightful weather, but expressed in a more convenient form.

17 Weather Variable

Chapter 17

Britain enjoys, if that is the word, the most variable weather in the world. Indeed, the only certain thing about British weather is that it is unlikely to remain the same for very long. Occasionally it forgets itself and remains settled for a few days. Usually, however, it changes from day to day and sometimes, when at its most fickle, from hour to hour. It is hardly surprising that the weather is such a common topic of conversation. This chapter suggests reasons for its variability.

Air masses

The air that affects Britain comes from varied sources. Air that originates in sub-tropical or polar latitudes, over land or sea areas, extends over Britain bringing conditions of temperature and humidity alien to our latitudes. Fig 205 shows the approximate position of the source areas. Over the great extent of North Africa for instance, temperatures are high and humidity is low. In this region there is a great expanse of air of uniform conditions – an *air mass*. In similar latitudes, but over the Atlantic Ocean, the air over a vast area again has uniform characteristics. Here is located

Fig 205

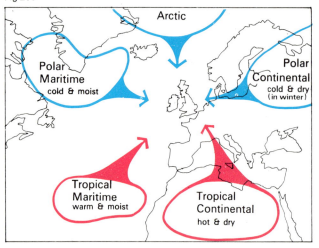

another air mass, the Tropical Maritime. Formed over the sub-tropical ocean, temperature and humidity are both high. Three other air masses are located in high latitudes. Polar Maritime air, originating over northern oceans, is moist as well as cold. Polar Continental is even colder, but dry, because it originates over a land mass. The source of the Arctic air mass is the frozen wastes of the polar ocean.

These air masses are never stationary for very long. They are in constant conflict like five armies sharing a battlefield. They advance in an attempt to dominate and control the no-man's land between them, and only retreat in the face of greater strength. As Fig 205 shows Britain is situated in the middle of the battlefield. Should the Tropical Continental air mass advance northwards, for instance, it will bring with it African air, cooled somewhat on the journey it is true, but still warm enough for Britain to bask in heat-wave conditions. We generally welcome the infrequent advances of Tropical Continental, but only plumbers and ice-skaters appreciate the occasional winter incursion of Polar Continental with its biting winds and freezing temperatures. For most of the year the maritime air masses are strong enough to dominate our weather. Fortunately the Polar Maritime air is warmed some-what by its passage over the ocean. Britain's equable climate is the result of this domination by air from over the oceans.

And so the battle rages. In January, for example, when a spell of cold showery weather gives way to a period of spring-like temperatures we know that, for a time at least, Tropical Maritime has overcome the resistance of Polar Maritime. When cool, showery weather spoils our summer holiday we can, at least, cheer a victory for Polar Maritime.

Depressions

As they jostle for position, air masses are constantly in contact. They do keep themselves very much to themselves, however, and there is very little mixing of air where they meet. The line separating two air masses is known as a *front*. The most important of these is the Polar Front where Polar and Tropical Maritime air masses meet. It is important because it is here that depressions are born. These low pressure systems, identified in Chapter 16, sweep eastwards with the prevailing westerly winds and are responsible for much of the variability of our weather.

The life story of a typical depression is told by the diagrams below. Fig 206 illustrates a typical situation on the Polar Front, probably somewhere out in the north-west Atlantic. The Polar Front separates the cold, dense polar air from the Tropical Maritime which is much warmer, lighter and contains much moisture. Wind directions correspond to those of the normal planetary winds, and at the front they are roughly parallel, but blowing in opposite directions. It must be stressed that the atmosphere has three dimensions, and that the Polar Front is not just a line on the surface but extends upwards for thousands of metres. Nor is the front a straight line, and in the diagram a slight bulge is shown at A. Note that at this point the warm south-westerly wind meets the cold air at an angle.

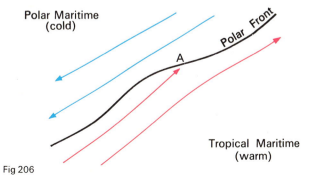

Fig 206

In Fig 207 the depression has grown older and the bulge has grown bigger. It is now a large salient of tropical air, known as the warm sector, protruding into the Polar

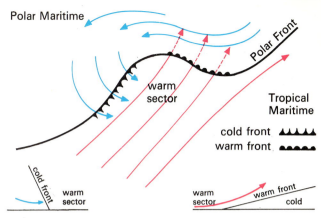

Fig 207

Front. The north-east winds bend round this warm sector and an anti-clockwise air circulation is building up. At this stage, two distinct parts of the Polar Front may be recognised. They are known as the warm and cold fronts and are identified on the diagram by the appropriate symbols. The sections included in Fig 207 show that the warm front is a gently sloping surface, whereas the cold front is much steeper. It is important to appreciate what is happening at these fronts. At the warm front the warmer and hence lighter, tropical air rises steadily over the heavy, polar air. It rises into areas of lower pressure and hence cools by expansion to give a lengthy period of rain. At the cold front the heavy, cold air, hugging the earth's surface, pushes under the warm air and lifts it up almost vertically. Here again we have rain, but this time it is usually in the form of heavy showers.

Of the depression's two fronts, the cold travels faster and gradually catches up with the warm. When the fronts meet *occlusion* is said to take place. Cold air meets cold air at the surface, and the warm sector is aloft. Occlusion starts at the apex of the salient and extends along the warm front. Fig 208 shows an early stage in the process and Fig 209 represents the situation a few hours later. The symbol for an occluded front indicates the presence of warm air at higher levels.

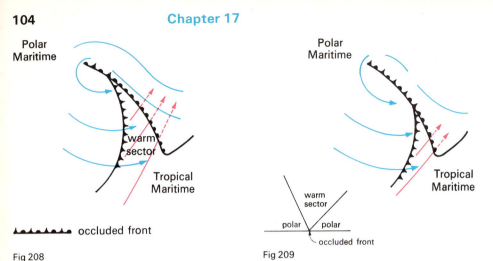

Polar
Maritime

warm
sector

Tropical
Maritime

▲▲▲▲▲▲▲ occluded front

Fig 208

warm
sector

polar   polar

occluded front

Fig 209

Polar
Maritime

Tropical
Maritime

Fig 211

Gradually, as the depression grows old the cold front overtakes the warm over the whole of its length and occlusion is completed. The whole of the warm sector has been lifted off the surface to disperse slowly in the upper atmosphere as the depression quietly passes away.

The sequence of weather in a depression

The depressions born on the Polar Front move eastwards with the prevailing winds. They pass over the British Isles at all seasons, but are most common in the winter half of the year. Fig 210 shows a typical depression as it appears on the weather map. This whirling mass of air,

Fig 210

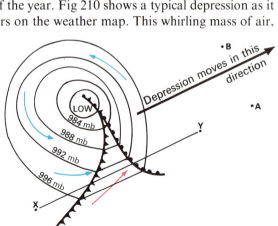

•B

Depression moves in this direction

•A

LOW
984 mb
988 mb
992 mb
996 mb

Y

X

perhaps 2000 kilometres across, moving in the direction shown by the arrow, will pass over an observer at A. It may take two or three days to pass and, for the observer, its passage will be marked by a sequence of pronounced changes in weather conditions.

This sequence is illustrated in Fig 212, which is a section through the depression along line XY. Imagine the diagram slipping slowly off the page to the right and you will get an idea of the movement of the depression over the observer at A. The first sign of the approaching depression is the appearance in the sky of faint whisps of cirrus cloud, Fig 211. This forms at about 10 000 metres and is composed of ice crystals. It thickens to veil the sky and the sun appears as a pale disc, often with a halo. Cloud becomes thicker and lower and eventually rain sets in. It persists for a period of hours, while the cloud cover develops into dark, low nimbostratus.

With the passage of the warm front, the observer, now in the warm sector, will notice several changes. The persistent rain will stop, cloud cover is lighter, temperatures are higher and the wind veers to the south-west. A barometer will reveal that pressure is no longer falling. The approach of the cold front is heralded by an increase of cloud. Tall, towering cumulus clouds give rain in short, heavy showers. After the cold front has

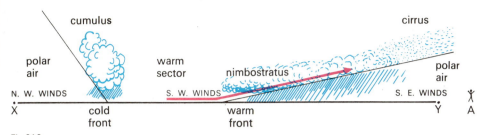

cumulus
cirrus

polar
air

warm
sector

nimbostratus

polar
air

N. W. WINDS
S. W. WINDS
S. E. WINDS

X
cold
front
warm
front
Y
A

Fig 212

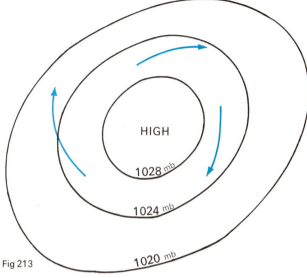

HIGH

1028 mb

1024 mb

1020 mb

Fig 213

passed, the observer is again in polar air. Temperatures are lower, pressure higher and the wind direction has changed yet again.

Consult Fig 210 once more. Note that it will be the occluded front that passes over an observer at B. As the tropical air is here above the surface, the observer will not experience the weather associated with the warm sector. Thus the weather recorded in the passage of a depression depends on the observer's position in relation to its line of movement, which often shows frequent changes of direction. As the speed of a depression may also vary considerably, we can perhaps appreciate some of the problems facing the meteorologist as he prepares his weather forecast.

Anticyclones

On the weather maps in Chapter 16 we identified another significant pattern in the isobars. This was the anticyclone or *high*. The typical pattern is shown in Fig 213. Note that the highest pressure is in the centre. The pressure gradient is gentle and winds are light. They blow outwards from the centre in a clockwise circulation.

Highs are associated with more than one type of weather. Sometimes they give dull cloudy conditions, but more characteristically the atmosphere is clear. In this case, the actual weather varies according to season. In summer there is little to impede insolation. Temperatures are high and a spell of fine, sunny weather is enjoyed. In winter, however, with no clouds to restrict the loss of heat from the earth, temperatures fall rapidly during the night and frost and fog are common before morning.

Finally, turn again to the weather maps on page 100. There, portrayed in the meteorologist's shorthand, are examples of actual depressions and anticyclones. Careful study will be rewarded by a greater understanding of these features which are responsible for our notoriously unpredictable weather.

Some Examination Questions

1 By reference to examples, show the difference between igneous, sedimentary and metamorphic rocks. (*Oxford and Cambridge Schools Examination Board*)

2 Describe and explain the physical features associated with rift valleys. (*Oxford Delegacy of Local Examinations*)

3 Describe, with examples and diagrams, the origin and characteristic features of (a) young fold mountains (b) block mountains. (*Welsh Joint Education Committee*)

4 Describe and explain, with the use of examples, the differences between fold mountains and other types of mountain. (*University of London School Examinations Council*)

5 (a) In **each** of **two** continents name (i) a volcano (ii) a rift valley. (You need not select (i) and (ii) from the same continent.)
(b) With the aid of diagrams, describe the physical features and method of formation of **one** of the volcanoes and **one** of the rift valleys you have named in (a). (*University of Cambridge Local Examinations Syndicate*)

6 Expand and explain **three** of the following statements:
(a) Coastlines are affected by changes in sea level.
(b) Currents on the west coast of Africa, south of the Equator, flow in a different direction from those on the east coast.
(c) The water off the coast of Labrador is colder than that off the coast of British Columbia.
(d) The ocean floor differs considerably in depth. (*Oxford and Cambridge Schools Examination Board*)

7 With the aid of diagrams, explain how **two** of the following have been formed: V-shaped valley; flood plain; river terrace; ria; delta. (*Oxford Delegacy of Local Examinations*)

8 Waterfalls, gorges, flood plains and deltas may occur in river courses. Choose **two** of these and for **each** (a) locate an example by means of a sketch map; (b) describe the physical characteristics of the example and the way in which it has formed. (*Associated Examining Board for the General Certificate of Education*)

9 With the aid of sketch maps and diagrams, describe **three** different landforms typical of the plain course of a large river and explain how **one** of your chosen landforms probably developed. (*Welsh Joint Education Committee*)

10 With the aid of diagrams, give an explanatory account of **three** characteristic physical features of the valley of a river which rises in mountain country, flows through an area of foothills and crosses a flood plain on its way through a delta to the sea. (*Joint Matriculation Board*)

11 Choose **three** of the following features: waterfall; levee; river terrace; delta; gorge. For **each** chosen feature (a) name a suitable example; (b) describe with the aid of diagrams the probable method of formation; (c) suggest ways in which the feature may be used by man. (*Oxford and Cambridge Schools Examination Board*)

12 (a) With the aid of diagrams, describe the physical features of a limestone region.
(b) Select any three of these features, and explain how they may have been formed.
(c) Name and locate **one** large limestone region. (*University of Cambridge Local Examinations Syndicate*)

13 Describe **three** features resulting from the glaciation of an upland region. (*Oxford Delegacy of Local Examinations*)

14 With the aid of diagrams, describe the appearance and mode of formation of **three** of the following: cirque (cwm or corrie); fjord; hanging valley; ribbon lake; terminal moraine. (*University of London School Examinations Council*)

15 In a glaciated area the following features may be found: cirque or corrie; hanging valley; pyramidal peak; terminal moraine; truncated spur; U-shaped valley.
(a) State briefly what you understand by **five** of the terms.
(b) With the aid of diagrams, show how any **two** are formed. (*University of London School Examinations Council*)

16 With the aid of diagrams and by reference to examples describe the effects of glaciation on the relief and drainage of **either** highland **or** lowland areas. (*Oxford and Cambridge Schools Examination Board*)

17 With the aid of diagrams, describe **three** important features of a hot desert landscape. (*Oxford and Cambridge Schools Examination Board*)

18 Naming an example in each case, give an account of **three** features of a coastline which have been formed by the action of the sea. (*Welsh Joint Education Committee*)

19 The present character of actual coastlines may have been due to one of the following causes:
(a) the destructive action of the sea;
(b) the constructive action of the sea;
(c) submergence of a land area.
For **each** of these name and locate **one** example and with the help of sketches describe its physical characteristics. (*Associated Examining Board for the General Certificate of Education*)

20 Choose **three** of the following: a ria coastline; a haff coastline; a fjord; a coral reef. For **each** (i) state the location of a particular example; (ii) explain and illustrate by diagrams how such a feature has been formed. (*Joint Matriculation Board*)

21 Choose **three** of the following landforms: fjord; barrier reef; delta; lagoon. For **each** (i) with the aid of diagrams describe its main features and suggest how the landform may have been formed; (ii) name an example and locate it by means of a sketch map. (*University of Cambridge Local Examinations Syndicate*)

22 (a) Describe three ways in which lakes can be formed.
(b) Give examples to show how lakes may be of use to man.
(*University of London School Examinations Council*)

23 Choose **three** of the following: a valley of a river in its middle course; a glacial valley; a rift valley; a valley in a desert region; a gorge in limestone country. For **each,** with the help of annotated diagrams:
(a) locate and name an actual example;
(b) describe its physical appearance;
(c) describe the stages by which it has developed to its present form. (*Associated Examining Board for the General Certificate of Education*)

24 (a) Describe the ways in which **either** a glacier **or** a river erodes its valley.
(b) With the aid of diagrams **or** contoured sketch maps, describe the chief contrasts between the features of a glaciated valley and those of a river valley. (*University of London School Examinations Council*)

25 Describe, and explain the mode of formation of **three** of the following features: cwm (corrie); a sea cliff; desert dunes; a limestone gorge. (*Welsh Joint Education Committee*)

26 Choose **three** of the following geographical features: a spring line; a ria; an artesian basin; a glaciated valley; an atoll. For each feature chosen (i) name and locate an actual example; (ii) describe and account for its chief characteristics. (*Welsh Joint Education Committee*)

27 Describe the physical features of each of **three** of the following: a glaciated upland; the plain course of a great river; a sand desert; an area of volcanic activity; a limestone upland. Illustrate your answer with sketch maps or diagrams. (*Joint Matriculation Board*)

28 Choose **three** of the following: an atoll; a corrie (cirque or cwm); a delta; a river terrace; an underground water course.
For **each** (i) state the location of a particular example; (ii) explain and illustrate by diagrams how such a feature has been formed. (*Joint Matriculation Board*)

29 Select **three** of the following: a coral reef; fjord; a canyon; a rift valley; an underground watercourse.
For **each** (i) state the location of a particular example; (ii) explain how such a feature has been formed, using diagrams to illustrate your answer. (*Joint Matriculation Board*)

30 Landforms may be created by the erosive action of rivers, glaciers and the sea. Describe and explain the formation of **one** landform attributable to each of these agents. Illustrate your answers by means of maps, sketches and diagrams. (*Welsh Joint Education Committee*)

31 Draw labelled diagrams to illustrate **two** of the following:
(a) wind as an agent of erosion;
(b) the effect of ground water in limestone regions;
(c) the formation of fjords. (*University of London School Examinations Council*)

32 (a) Name **one** example of **each** of the following: fold mountains; block mountains; gorge; waterfall.
(b) For any **two** given in (a) describe **each** and say how it may have been formed. (*University of Cambridge Local Examinations Syndicate*)

33 Describe the physical characteristics and mode of formation of **three** of the following: arête; crag-and-tail; ox-bow lake; barchan; sea-stack. (*Associated Examining Board for the General Certificate of Education*)

34 Select **three** of the following physical features: hanging valley; lava plateau; ribbon lakes; arête; fault scarp; outwash plain. For **each** of the three chosen: (a) name an example; (b) with the aid of diagrams (i) describe its appearance; (ii) comment on the factors that have affected the landscape of which it forms part. (*Oxford and Cambridge Schools Examination Board*)

35 Choose **three** of the following physical features: ria; horst (block mountain); delta; barrier reef; cirque (corrie); saltpan. For **each** of those you have chosen, with the help of sketch maps or diagrams:
(a) name and locate a precise example;
(b) describe its appearance;
(c) explain its mode of formation. (*Associated Examining Board for the General Certificate of Education*)

36 With the aid of diagrams and examples, explain **three** of the following: convectional rainfall; equable climate; inland drainage; monsoon winds; ria coast; spring-line settlement; waterfall. (*University of London School Examinations Council*)

37 With the aid of diagrams:
(a) distinguish between the effects of the rotation and the revolution of the earth;
(b) explain why the length of day and night varies in different parts of the world;
(c) explain why the time in Western Australia is 20.00 when it is noon in Britain. (*University of London School Examinations Council*)

38 With the aid of diagrams answer the following:
(a) where can the sun be seen continually for 24 hours or more?
(b) where is the sun never seen for more than twelve hours in a day?
(c) why are the seasons reversed in the northern and southern hemispheres of the world? (*Oxford Delegacy of Local Examinations*)

39 (a) What is understood by the term 'Tropic of Cancer'?
(b) Explain why the length of daylight at a place on latitude 50°N varies with the seasons of the year.
(c) State and explain fully the difference in local time at Rome (41°54'N, 12°28'E) and Santos (23°58'S, 46°20'W). (*Associated Examining Board for the General Certificate of Education*)

40 Answer the following, illustrating your answers with diagrams:
(a) The table below gives the approximate length of daylight on 21st June:

Lat. 0°	12 hours
Lat. 23½°N	13½ hours
Lat. 66½°N	24 hours

Explain why the length of daylight varies with latitude in this way.
(b) The position of a ship is given as 50°N, 25°W. How does this information enable the position of the ship to be known?
(c) In London I was eating my lunch at 14.00, but at the same moment my friend in Chicago was having breakfast. Explain how this statement can be true. (*University of Cambridge Local Examinations Syndicate*)

41 With the aid of diagrams, explain:
(a) the term 'equinox';
(b) why a day is lost when one travels round the world from east to west;
(c) why the Polar lands beyond the Arctic Circle have prolonged daylight in summer. (*Associated Examining Board for the General Certificate of Education*)

42 (i) With the aid of clearly labelled diagrams, explain why: the period of daylight in December in the British Isles is less than that of June;
(ii) the extreme north of Norway is referred to as the 'Land of the Midnight Sun';
(iii) when it is midday in London it is 06.00 in New Orleans, U.S.A. (*Welsh Joint Education Committee*)

43 Explain, with the help of diagrams, the meaning of the following terms: Greenwich meridian; equinox; time zones. (*Welsh Joint Education Committee*)

44 (a) Explain the meaning of the terms 'equinox' and 'solstice'.
(b) With the aid of diagrams, show how they are related to the movements of the earth. (*University of Cambridge Local Examinations Syndicate*)

45 With the aid of clearly labelled diagrams, explain:
(a) why the result of a rugby football match played in Brisbane, Australia, on Saturday, 4th June 1966, in the afternoon, was announced in Cardiff on the morning of the same day;
(b) why the altitude of the midday sun is at its lowest in New Zealand on 21st June;
(c) why there are several weeks of total darkness in Polar regions each year. (*Welsh Joint Education Committee*)

46 With the aid of diagrams, explain the need for standard time zones and for the International Date Line. (*University of London School Examinations Council*)

47 (a) Study the following extract from the timetable of a B.O.A.C flight from Sydney to San Francisco and, by reference to diagrams, explain how an aircraft leaving Sydney on a Tuesday can arrive at San Francisco earlier than it took off.

Sydney	dep	Tuesday	19.00 hours
Fiji	arr	Wednesday	00.40 hours
Honolulu	arr	Tuesday	09.25 hours
	dep	Tuesday	10.40 hours
San Francisco	arr	Tuesday	18.20 hours

(b) With the aid of a diagram, explain why the lengths of daylight vary little throughout the year on the Equator. (*Oxford and Cambridge Schools Examination Board*)

48 With the aid of diagrams, explain why:
(a) opencast mining in northern Sweden can proceed during the winter months only with the aid of floodlights;
(b) the warm sector is the part of a depression most likely to produce rain;
(c) a stationary anticyclone in winter may bring dense fog to the London area. (*Oxford and Cambridge Schools Examination Board*)

49 Explain, using diagrams, the geographical factors affecting temperature. (*University of London School Examinations Council*)

50 (a) What instruments are essential at a weather station and what information do they provide?
(b) Describe briefly how the following climatic statistics are compiled: average annual rainfall; mean monthly temperature. (*University of London School Examinations Syndicate*)

51 (a) List the main instruments used in a school weather station.
(b) Draw labelled diagrams of **two** of them.
(c) Describe **two** interesting series of weather observations you have recorded yourself. (*University of London School Examinations Council*)

52 (a) Name and describe the instrument used for measuring precipitation at a school weather station.
(b) State what you understand by 'mean annual rainfall' and explain what steps must be taken to calculate it for a particular station. (*Associated Examining Board for the General Certificate of Education*)

53 (a) Briefly describe the instruments used to obtain measurements of daily rainfall, temperature and relative humidity, explaining the precautions taken to obtain accurate readings.
(b) Show how weather observations are plotted on a typical weather map.
(c) Explain how the mean monthly temperature for a place may be calculated. (*Welsh Joint Education Committee*)

54 Describe how air temperature, humidity and winds are recorded at a school weather station. Explain how this information may be mapped and used in the geography room. (*Oxford and Cambridge Schools Examination Board*)

55 (a) What is meant by the term 'precipitation'?
(b) Draw an annotated diagram to illustrate a rain-gauge.
(c) What is understood by 'mean annual rainfall'? Explain how this is calculated.
(d) Name the **three** types of rainfall. In the case of any **one** explain the processes involved. (*Associated Examining Board for the General Certificate of Education*)

56 With the aid of diagrams, and by reference to actual examples, describe **three** of the following, and state clearly how **each** of the **three** has been caused:
(a) land and sea breezes; (b) a rain-shadow area; (c) föhn (chinook) winds; (d) hurricane (typhoons). (*University of Cambridge Local Examinations Syndicate*)

57 What is a föhn or chinook wind? Draw an annotated diagram to illustrate how it occurs and its effects.
What are the causes of a persistent fog? (*Associated Examining Board for the General Certificate of Education*)

58 Show how symbols are used to explain weather conditions in a synoptic chart. (*University of London School Examinations Council*)

59 Describe and account for the weather conditions associated with the passage of a depression over the British Isles. (*Welsh Joint Education Committee*)

60 (a) What thermometers would you provide for a school meteorological station? Explain your choice.
(b) Assume that a small depression, with its centre to the north of you, passes over your school. Give an explanatory account of the changes likely in (i) wind direction; (ii) barometric pressure; (iii) temperature; (iv) rainfall. (*Oxford and Cambridge Schools Examination Board*)

Index